QUARKS
Selected Reprints

Edited by O.W. Greenberg

Department of Physics and Astronomy
University of Maryland, College Park

published by
American Association of Physics Teachers

Published by
American Association of Physics Teachers
Publications Department
5110 Roanoke Place, Suite 101
College Park, MD 20740, U.S.A.

Cover design: Karen S. Garvin

ISBN #0-917853-17-2

Contents

I. Background

II. Quark Degrees of Freedom

III. Quarks as Static Constituents of Hadrons

IV. Quarks as Partons

V. Quantum Chromodynamics and Permanent Quark and Color Confinement

VI. Unified Theories

VII. The Next Layer of the Onion? Possible Quark Compositeness

VIII. Astrophysical and Cosmological Implications of Quarks

Resource Letter Q-1: Quarks

O. W. Greenberg
Center for Theoretical Physics, Department of Physics and Astronomy, University of Maryland, College Park, Maryland 20742

(Received 5 March 1982; accepted for publication 22 March 1982)

Quarks as fundamental constituents of hadrons play a central role in elementary particle physics. We give an annotated bibliography of references to quarks and related topics in elementary particle physics, as well as to the role of quarks in areas outside elementary particle physics, such as nuclear physics, and astrophysics and cosmology. We label references E (elementary), I (intermediate), and A (advanced) to guide the reader. Articles selected for incorporation in a reprint volume (to be published separately by the American Association of Physics Teachers) are indicated by (*). A short list of particularly helpful elementary and intermediate references is indicated by (⋆).

PREFACE

The subject "quarks" has a large literature. In this Resource Letter, I have tried to gather the original references to a given development, important subsequent references, review articles, reprint collections, and, perhaps most important, elementary references which will be accessible to interested people whose field of specialization is outside elemenary particle physics. It hardly needs to be said that all references which I cite, especially review articles, contain further citations. It may also be superfluous to mention that more citations, as well as citations to work yet to be done, appear in present (and will appear in future) volumes of the conference proceedings, summer and winter school proceedings, workshop proceedings, journals, etc., cited here, as well as in computerized citation systems. The Science Citation Index can be used to find later articles on a given subject by finding the citations to an earlier article on that subject. The Particle Data Group publishes an updated "Review of Particle Properties." in Rev. Mod. Phys. or Phys. Lett. every other year.

I have given citations to the English translations of Soviet publications, because these are likely to be more useful to readers of this Resource Letter than citations to the original Russian language articles. I have cited laboratory publications from Brookhaven National Laboratory, Centre European Recherche Nucleaire, Deutsches Electronen–Synchrotron, Fermi National Accelerator Laboratory, Lawrence Berkeley Laboratory and Stanford Linear Accelerator Center as BNL, CERN, DESY, FERMILAB, LBL, and SLAC, respectively.

I. INTRODUCTION

An important result of the study of matter at the microscopic level has been the discovery of layers of substructure: molecules are the smallest objects having specific chemical properties; atoms are the objects whose combinations give molecules; nuclei and electrons are the objects which form atoms; protons and neutrons are the objects which form nuclei. By 1940, the elementary particles known were the proton, neutron, and electron just mentioned, the light quantum or photon, the positron (antiparticle of the electron), the electron neutrino and antineutrino and the muon (although the neutrino and antineutrino had not been detected directly, and the muon was thought to be the pion predicted by Yukawa). Charged pions were first produced in accelerator experiments at the Berkeley 184-in. synchrocyclotron, as was the neutral pion. Kaons, and the lambda (Λ^0), sigma ($\Sigma^{+,0,-}$), and xi ($\Xi^{0,-}$) hyperons were found in the late 1940s and the 1950s. The apparent clash between the production and decay mechanisms of these particles was resolved by the strangeness classification introduced in

1. **"Isotopic Spin and New Unstable Particles,"** M. Gell-Mann, Phys. Rev. **92**, 833–834 (1953). (A)

*2. **"The Interpretation of the New Particles as Displaced Charge Multiplets,"** M. Gell-Mann, Nuovo Cimento **4**, Suppl. No. 2, 848–866 (1956). (A)

3. **"Charge Independence for U-particles,"** T. Nakano and K. Nishijima, Prog. Theor. Phys. **10**, 581–582 (1953). (A)

4. **"On the Theory of Hyperons and Heavy Mesons,"** K. Nishijima, Fortschr. Phys. **4**, 519–559 (1956). (A)

The separation of the particles into the broad categories of baryons (including protons, neutrons, hyperons, and their excited states), mesons (including pions, kaons, and their excited states), the photon, and leptons (now including the electron, muon, tau, and their neutrinos), together with the antiparticles of baryons and leptons is discussed in Ref. 2. The photon is its own antiparticle. The mesons as a set are their own antiparticles; for example, the π^+ and π^- are each other's antiparticle; and the π^0 is its own antiparticle. Baryons and mesons together are called hadrons. The $SU(3)_{\rm flavor}$ or $SU(3)_f$ (originally called unitary symmetry or "the eightfold way") description of hadrons was introduced in

5. **"The Eightfold Way: A Theory of Strong Interaction Symmetry,"** M. Gell-Mann, Caltech preprint CTSL-20 (1961), reprinted in Ref. 7 below. (A)

6. **"Derivation of Strong Interactions from a Gauge Invariance,"** Y. Ne'eman, Nucl. Phys. **26**, 222–229 (1961). (A)

The status of the $SU(3)_f$ description just before the introduction of quarks is given in

7. **The Eightfold Way**, M. Gell-Mann and Y. Ne'eman (Benjamin, New York, 1964). (A)

which contains reprints of important articles. Although the fundamental triplet representation of the group SU(3) (references to group theory are given in the next paragraph) was not used in the flavor classification, the triplet representation was used as a mathematical device by Gell-Mann in Ref. 5 in constructing the octet representation to which observed particles were assigned. An interesting attempt to use the group U(3) to describe hadrons by introducing weights $(\lambda_1,\lambda_2,\lambda_3)$ which are related to the differences of the number of quarks and the number of antiquarks of a given flavor (flavor will be discussed in Sec. II) by

$$n_u - n_{\bar u} = (2\lambda_1 + 2\lambda_2 - \lambda_3)/3,$$

$$n_d - n_{\bar d} = (2\lambda_1 - \lambda_2 + 2\lambda_3)/3\,,$$

and

$$n_s - n_s = (-\lambda_1 + 2\lambda_2 + 2\lambda_3)/3\,.$$

was made in

8. **"Baryon Charge and R-Inversion in the Octet Model,"** H. Goldberg and Y. Ne'eman, Nuovo Cimento **27**, 1–5 (1963). (A)

Group theory, in particular the theory of representations of Lie groups, is a very useful mathematical tool for the description of symmetries and the classification of particles. Relevant group theory is discussed in

9. **Lie Groups for Pedestrians**, H. J. Lipkin (North-Holland, Amsterdam, 1965). (I)

10. **"The Octet Model and its Clebsch–Gordon Coefficients,"** J. J. de Swart, Rev. Mod. Phys. **35**, 916–939 (1963). (A)

11. **"Unitary Groups: Representations and Decompositions,"** C. Itzykson and M. Nauenberg, Rev. Mod. Phys. **38**, 95–120 (1966). (A)

12. **Group Theory**, M. Hamermesh (Addison-Wesley, Reading, MA, 1962). (A)

13. **Introduction to Unitary Symmetry**, P. Carruthers (Interscience, New York, 1966). (A)

A model of the pion as a bound state of a nucleon and antinucleon was constructed in

*14. **"Are Mesons Elementary Particles?,"** E. Fermi and C. N. Yang, Phys. Rev. **76**, 1739–1743 (1949). (A)

This model is a prototype of later attempts to construct objects on the level of baryons and mesons in terms of each other or of more elementary objects.

The decisive step of introducing quarks was taken by

15. **"A Schematic Model of Baryons and Mesons," M. Gell-Mann, Phys. Lett. **8**, 214–215 (1964). (A)

16. G. Zweig, CERN 8182/TH. 401, 1964 (unpublished). (A)

17. **"An SU(3) Model for Strong Interaction Symmetry and Its Breaking,"** G. Zweig, CERN 8419/TH. 412, 1964 (unpublished), reprinted in Ref. 30 below (A)

The name "quark," chosen by Gell-Mann, comes from

18. **Finnegan's Wake**, J. Joyce (Viking, New York, 1939), p. 383. (A)

Elementary introductions to the quark model are given in

19. **Subatomic Physics**, H. Frauenfelder and E. M. Henley (Prentice-Hall, Englewood Cliffs, NJ, 1974). (E)

20. **The Physics of Elementary Particles**, L. J. Tassie (Wiley, New York, 1973). (E)

21. **Fundamentals of Elementary Particle Physics**, M. J. Longo (McGraw-Hill, New York, 1973). (E)

Zweig describes his path to the invention of the quark model in

22. **"Origins of the Quark Model,"** G. Zweig, in *Baryon 1980, Proc. IVth International Conference on Baryon Resonances, Toronto, 1980*, edited by N. Isgur (University of Toronto, Toronto, 1981), pp. 439–459. (I)

More sophisticated surveys of the quark model are given in

23. **Models of Elementary Particles**, B. T. Feld (Blaisdell, Waltham, MA, 1969). (I)

24. **The Quark Model**, J. J. Kokkedee (Benjamin, New York, 1969). (A)

25. **"Quarks for Pedestrians,"** H. J. Lipkin, Phys. Rep. C **8**, 173–268 (1973). (A)

26. **"The Quark Model,"** A. W. Hendry and D. B. Lichtenberg, Rep. Prog. Phys. **41**, 1707–1780 (1978). (A)

27. **"Quarks,"** O. W. Greenberg, Ann. Rev. Nucl. Part. Sci. **28**, 327–386 (1978). (A)

28. **An Introduction to Quarks and Partons**, F. E. Close (Academic, New York, 1979). (A)

29. **"Quark Models,"** J. L. Rosner, in *Techniques and Concepts of High Energy Physics*, NATO Advanced Studies Institute Series, edited by T. Ferbel (Plenum, New York, 1981), pp. 1–141. (A)

30. **Developments in the Quark Theory of Hadrons**, edited by D. B. Lichtenberg and S. P. Rosen (Hadronic, Nonantum, 1980). (A)

31. **"Hadron Spectra and Quarks,"** S. Gasiorowicz and J. L. Rosner, Am. J. Phys. **49**, 954–984 (1981). (A)

References 24 and 30 contain reprints of important articles.

Quarks are used in several ways in elementary particle physics: as constituents of static hadrons, as fields from which currents and charges can be constructed, and as partons in the description of high-energy collisions.

Elementary particle physics (including quark physics) is discussed in an elementary way in

32. **What is the World Made of? Atoms, Leptons, Quarks, and Other Tantalizing Particles**, G. Feinberg, (Anchor/Doubleday, Garden City, NY, 1977). (E)

33. **The Forces of Nature**, P. C. W. Davies (Cambridge University, Cambridge, 1979). (E)

34. **The Particle Play, An Account of the Ultimate Constituents of Matter** J. C. Polkinghorne (Freeman, San Francisco, 1979). (E)

35. **"Elementary Particles in Physics,"** S. Gasiorowicz, in *Encyclopedia of Physics*, edited by R. G. Lerner and G. L. Trigg (Addison-Wesley, Reading, MA, 1981), pp. 270–283. (E)

36. **From Atoms to Quarks, An Introduction to the Strange World of Particle Physics**, J. S. Trefil (Scribners, New York, 1980). (E)

A number of relevant references are cited in an earlier Resource Letter:

37. **"Resoure Letter NP-1: New Particles,"** J. L. Rosner, Am. J. Phys. **48**, 90–103 (1980). (E)

An interesting overview of theoretical physics is given in

*38. **"Is the End in Sight for Theoretical Physics,"** S. Hawking, inaugural lecture as incumbent of the Lucasian Chair of Mathematics, Univ. of Cambridge, in CERN Courier **21**, 3–8 and 71–74 (1981). (I)

A film which discusses quarks and elementary particle physics in an elementary way is

39. **The Hunting of the Quark** (Time-Life, New York, 1974). (E)

Elementary particle physics (including quark physics) is discussed at an intermediate and advanced level in

40. **Elementary Particles,** W. I. Frazer (Prentice-Hall, Englewood Cliffs, NJ, 1966). (I)

41. **Elementary Particle Physics,** S. Gasiorowicz (Wiley, New York, 1966). (A)

42. **High Energy Hadron Physics,** M. Perl (Wiley, New York, 1974). (A)

II. QUARK DEGREES OF FREEDOM

Quarks carry three types of intrinsic degrees of freedom, intrinsic angular momentum (or spin), flavor, and color. Brief elementary descriptions of spin can be found in many elementary physics texts, such as Ref. 19, pp. 68–71. Spin is discussed at an intermediate level in

43., **The Feynman Lectures on Physics, Vol. 3, Quantum Mechanics,** R. P. Feynman, R. B. Leighton, and M. Sands (Addison-Wesley, Reading, MA, 1965), pp. 6-1–6-14. (I)

The role of spin in atomic physics is extensively discussed in

44. **Atomic spectra and Atomic Structure,** G. Herzberg (Dover, New York, 1944). (A)

45. **Group Theory and its Application to the Quantum Mechanics of Atomic Spectra,** E. P. Wigner (Academic, New York, 1959). (A)

46. **Quantum Theory of Angular Momentum, A Collection of Reprints and Original Papers,** edited by L. C. Biedenharn and H. Van Dam (Academic, New York, 1965). (A)

The choice of spin $\frac{1}{2}$ for quarks made in Refs. 15–17, together with integer-valued orbital angular momenta (both in units of Planck's constant $h/2\pi = \hbar$) allows construction of all spin states of hadrons. The symmetric quark model of baryons (see the discussion of color, below) gives a detailed analysis of baryon spectroscopy which provides further evidence in the context of strong interactions that quarks have spin $\frac{1}{2}$,

47. **"Spin and Unitary-Spin Independence in a Paraquark Model of Baryons and Mesons," O. W. Greenberg, Phys. Rev. Lett. **13**, 598–602 (1964). (A)

*48. **"Symmetric Quark Model of Baryon Resonances,"** O. W. Greenberg and M. Resnikoff, Phys. Rev. **163**, 1844–1851 (1967). (A)

The Callan–Gross sum rule in the quark–parton model, which is verified experimentally to within 15%, provides further evidence that quarks have spin $\frac{1}{2}$ in the context of weak and electromagnetic interactions,

49. **"High Energy Electroproduction and the Constitution of the Electric Current," C. G. Callan and D. J. Gross, Phys. Rev. Lett. **22**, 156–159 (1969). (A)

The electric charge Q, hypercharge Y, and third component of the isospin I_3 of quarks, which are related by the Gell-Mann–Nishijima formula (Refs. 1–4).

$$Q = I_3 + \tfrac{1}{2} Y,$$

are associated with the flavor degree of freedom. These quantum numbers are discussed at an elementary level in Ref. 21, pp. 160–175. A charming discussion of these topics in the context of neutrino physics, is given in

*50. **"Fact and Fancy in Neutrino Physics,"** A. De Rujula, H. Georgi, S. L. Glashow, and H. R. Quinn, Rev. Mod. Phys. **46**, 391–407 (1974). (A)

Three flavors are required to construct the hadrons which were known before the discovery of charm. These are the up, down, and strange flavors associated with the u, d, and s quarks. The transformation of these quarks as a "unitary spin" triplet under an approximate SU(3) flavor or $SU(3)_f$ symmetry, for short, provides a rationale for the eightfold way, and leads to the observed irreducible multiplets: octets, decuplets, and singlets of hadrons. Mixing effects (see articles by Glashow and Sakurai in Ref. 5) lead to complications which obscure the fact that only these three multiplets occur. More complicated mixings occur in baryons, and are discussed in the references on baryon spectroscopy below. The electromagnetic current, expressed in terms of quark fields,

$$j^{\mu}_{\text{em}} = \tfrac{1}{3}(2\bar{u}\,\gamma^{\mu}u - \bar{d}\,\gamma^{\mu}d - \bar{s}\gamma^{\mu}s)$$

leads directly to the relative leptonic branching rates

$$\Gamma(\rho \to l^{+}l^{-}):\Gamma(\omega \to l^{+}l^{-}):\Gamma(\phi \to l^{+}l^{-}) = 9:1:2,$$

which are in good agreement with experiment. The Cabibbo rotated quark fields

$$\begin{pmatrix} u \\ d_C(\theta) \\ s_C(\theta) \end{pmatrix} = \begin{pmatrix} 1 & 0 & 0 \\ 0 & \cos\theta_C & \sin\theta_C \\ 0 & -\sin\theta_C & \cos\theta_C \end{pmatrix} \begin{pmatrix} u \\ d \\ s \end{pmatrix}$$

used to construct the charged ($\Delta Q = \pm 1$) Cabbibbo currents

$$j^{\mu}_{+} = \bar{u}\,\gamma^{\mu}(1+\gamma_5)d_C(\theta), \quad j^{\mu}_{-} = \bar{d}_C(\theta)\gamma^{\mu}(1+\gamma_5)u$$

lead to good agreement with experiment for charge-changing weak processes,

51. **"Unitary Symmetry and Leptonic Decays,"** N. Cabibbo, Phys. Rev. Lett. **10**, 531–533 (1963). (A)

Although several authors speculated on the possible existence of additional flavors of quarks, the first compelling theoretical argument for an additional flavor, charm, was given in

52. **"Weak Interactions with Lepton–Hadron Symmetry," S. L. Glashow, J. Illiopoulos, and L. Maiani, Phys. Rev. D **2**, 1285–1292 (1970). (A)

These authors introduced the celebrated GIM mechanism which uses the charm quark to cancel the otherwise unavoidable $|\Delta S| = 1, \Delta Q = 0$ currents (which would, if uncancelled, lead to certain K-meson decays at rates many orders of magnitude greater than experimental upper limits). The GIM weak currents

$$\begin{aligned} j_{+}{}^{\mu} &= \bar{u}_L\,\gamma^{\mu}d_L(\theta) + \bar{c}_L\,\gamma^{\mu}s_L(\theta) = (j_{-}{}^{\mu})^{\dagger} \\ j_0{}^{\mu} &= \bar{u}_L\,\gamma^{\mu}u_L + \bar{d}_L(\theta)\,\gamma^{\mu}\,d_L(\theta) \\ &\quad + \bar{c}_L\,\gamma^{\mu}c_L + \bar{s}_L(\theta)\gamma^{\mu}s_L(\theta) \\ &- \bar{u}_L\,\gamma^{\mu}\,u_L + \bar{d}_L\,\gamma^{\mu}\,d_L + \bar{c}_L\,\gamma^{\mu}c_L + \bar{s}_L\,\gamma^{\mu}s_L \end{aligned}$$

constructed from the left-handed weak doublets $[u_L, d_L(\theta)]$ and $[c_L, s_L(\theta)]$ where $q_L = (1+\gamma_5)q_C$, $\bar{q}_L = \bar{q}_C(1-\gamma_5)$ lead to the same $|\Delta Q| = 1$ transitions as the Cabibbo currents, without the unobserved $\Delta Q = 0$ transitions. A review of charm completed just before the discovery of the J/Ψ (a heavy vector meson made of a charm and an anticharm quark), and updated before publication to take account of this discovery is given in

53. **"Search for Charm,"** M. K. Gaillard, B. W. Lee, and J. L. Rosner, Rev. Mod. Phys. **47**, 277–310 (1975). (A)

An early discussion of the properties of the J/Ψ on the basis of quantum chromodynamics (see Sec. VI) is given in

54. **"Heavy Quarks and e^+e^- Annihilation,"** T. Appelquist and H. D. Politzer, Phys. Rev. Lett. **34**, 43-45 (1975). (A)

Charm is reviewed in

*55. **"Electron–Positron Annihilation and the New Particles,"** S. D. Drell, Sci. Am. **232** (6), 50–62 (1975). (A)

*56. **Quarks with Color and Flavor,"** S. L. Glashow, Sci. Am. **233** (4), 38–54 (1975). (E)

57. **"Fundamental Particles with Charm,"** R. F. Schwitters, Sci. Am. **237** (4), 56–70 (1977) (E)

58. **"Electron–positron Annihilation Above 2 GeV and the New Particles,"** G. J. Feldman and M. L. Perl, Phys. Rep. **19C**, 233–293 (1975). (A)

59. **"Charm and Beyond,"** T. Appelquist, R. M. Barnett, and K. Lane, Ann. Rev. Nucl. Part. Sci. **28**, 387–499 (1978). (A)

The fifth flavor of quark, b for bottom or beauty, was discovered without the benefit of prior explicit theoretical blessing in the Υ (upsilon) which is a $\bar{b}b$ resonance by

60. **"Observation of a Dimuon Resonance at 9.5 GeV in 400-GeV Proton-Nucleus Collisions,"** S. W. Herb, D. C. Hom, L. M. Lederman, J. C. Sens, H. D. Snyder, J. K. Yeh, J. A. Appel, B. C. Brown, C. N. Brown, W. R. Innes, K. Ueno, T. Yamaguchi, A. S. Ito, H. Jöstlein, D. M. Kaplan, and R. D. Kephart, Phys. Rev. Lett. **39**, 252–255 (1977). (A)

This quark, however, together with an as yet unfound sixth flavor of quark, t for top or truth, which, if it exists at all, must have a mass greater than 18 GeV (see Ref. 63 below), fits well into a theory of CP conservation, which requires at least six quark flavors due to

*61. **"CP-Violation in the Renormalizable Theory of Weak Interaction,"** M. Kobayashi and T. Maskawa, Prog. Theor. Phys. **49**, 652–657 (1973). (A)

The upsilon is discussed at an elementary level in

*62. **"The Upsilon Particle,"** L. M. Lederman, Sci. Am. **239** (4), 72–80 (1978). (E)

Both charm- and bottom-quark phenomena are reviewed in

63. **"New e^+e^- Physics,"** B. H. Wiik, in *High Energy Physics 1980, Proc. XX Int. Conf. on High Energy Physics, 1980*, edited by L. Durand and L. G. Pondrom (American Institute of Physics, New York, 1981), pp. 1379–1429. (A)

64. **"Phenomenology of New Particle Production,"** R. J. N. Phillips, *ibid.*, pp. 1471–1498. (A)

65. **"New Flavor Spectroscopy,"** K. Berkelman, *ibid.*, pp. 1500–1529. (A)

A review of nonrelativistic quantum-mechanical methods to study the variation of bound-state parameters with constituent mass and excitation energy, with applications to mesons composed of a quark and an antiquark, is given in

66. **"Quantum Mechanics with Applications to Quarkonium,"** C. Quigg and J. L. Rosner, Phys. Rep. **56**, 167–235 (1979). (A)

At present we do not know whether or not more quark flavors exist.

The present status of the dynamics of flavor is reviewed in

67. **"Quarks and Leptons,"** H. Harari, Phys. Rep. **42C**, 236–309 (1978).

68. **"Flavordynamics of Quarks and Leptons,"** H. Fritzsch and P. Minkowski, Phys. Rep. **73**, 67–173 (1981).

Color is a picturesque expression, having nothing to do with the colors that we see with our eyes, for a new degree of freedom carried by quarks. Although this new degree of freedom was introduced using para-Fermi statistics of order 3 (see below), it is simpler to describe using an exact $SU(3)_{color}$ or $SU(3)_c$ symmetry (see below). The quarks belong to the fundamental 3 representation of $SU(3)_c$. Thus color increases the number of quarks by a factor of three, i.e., there are three colors of u quarks, three colors of d quarks, etc. Hadrons are color singlets, so that hadrons are neutral with respect to color. If permanent color confinement (which includes permanent quark confinement) holds (see Sec. VI below), then color is a hidden degree of freedom in the sense that no isolated particles can carry (nonsinglet) color. Color was first introduced in order to resolve a paradox concerning the statistics of quarks. In the SU(6) theory discussed below, the ground-state baryons are composed of three quarks in a state which is symmetric under permutations of the degrees of freedom which were then known to be associated with each quark (spin, flavor, and the space degree of freedom). The Pauli exclusion principle for electrons states that at most one electron can occupy a given quantum state. Bohr used this principle to understand the periodic table of chemical elements. The impact of the exclusion principle on atomic spectroscopy and on chemistry are discussed in Refs. 44 and 45 for spectroscopy and in

69. **Introduction to Quantum Mechanics with Applications to Chemistry,** L. Pauling (McGraw-Hill, New York, 1935). (A)

for chemistry. The particles which obey the exclusion principle have "statistics" called Fermi–Dirac statistics. Particles which do not obey the exclusion principle, such as photons and pi mesons, have statistics called Bose–Einstein statistics. Empirically, Fermi–Dirac particles have spin $\frac{1}{2}$, $\frac{3}{2}$, $\frac{5}{2}$, etc. (i.e., odd-half-integer spin) and Bose–Einstein particles have spin 0, 1, 2, etc. (i.e., integer spin), as discussed in Ref. 19 and in

70. **Quantum Mechanics,** E. Merzbacher (Wiley, New York, 1970), pp. 515–516. (I)

The empirical connection between spin and statistics was shown to follow from the properties of free relativistic quantum fields by Pauli in the late 1930s and early 1940s. More recently this result has been established for interacting fields in a rigorous way in the context of axiomatic quantum field theory; it is discussed in

71. **"Exclusion Principle, Lorentz Group and Reflexion of Space-Time and Charge,"** W. Pauli, in *Niels Bohr and the Development of Physics*, edited by W. Pauli (McGraw-Hill, New York, 1955), pp. 30–51. (A)

72. **PCT Spin and Statistics, and All That,** R. F. Streater and A. S. Wightman (Benjamin, New York, 1964), pp. 146–161 (A)

The implication of this theorem for quarks is that since quarks have spin $\frac{1}{2}$ they should obey Fermi–Dirac statistics, and should have a wave function which is totally antisymmetric under permutations of the degree of freedom associated with each quark, which is precisely the reverse of the apparent situation in the SU(6) model. (Bose and Fermi "statistics" are associated with wave functions for identical particles which are symmetric and antisymmetric, respectively, under simultaneous permutations of all degrees of freedom associated with the particles.) This paradox was resolved by introducing para-Fermi statistics of order 3 for the quarks in Ref. 47. This allows three quarks to occupy the same flavor, spin, and space state. Additional evidence for three colors comes from the π^0 decay rate, which is nine times larger with color than without, from the ratio

$$R = (e^+e^- \to \text{hadrons})/(e^+e^- \to \mu^+\mu^-),$$

which is three times larger with color than without, and from the hadronic production of $\mu^+\mu^-$ pairs via the Drell–Yan mechanism (see Ref. 113, 114, and 115 below), which is one-third as large with color as without. The experimental evidence for color is discussed in Ref. 78 below. Reference 47 also introduced the shell model of baryons using effective symmetric statistics in the "visible" degrees of freedom. The explicit introduction of color using the group

Table I. Properties of quarks. The symbols are J^P = spin parity, Q/e = electric charge in units of the proton charge e, B = baryon number ($B = 1$ for the proton), $SU(3)_f$ = flavor SU(3) representation, I = total isospin, I_3 = third component of isospin, Y = hypercharge = $B + S$, S = strangeness, $SU(3)_c$ = color SU(3) representation.

Flavor	J^P	Q/e	B	$SU(3)_f$	I	I_3	S	Y	$SU(3)_c$
u	$\frac{1}{2}^+$	$\frac{2}{3}$	$\frac{1}{3}$	**3**	$\frac{1}{2}$	$\frac{1}{2}$	0	$\frac{1}{3}$	**3**
d	$\frac{1}{2}^+$	$-\frac{1}{3}$	$\frac{1}{3}$	**3**	$\frac{1}{2}$	$-\frac{1}{2}$	0	$\frac{1}{3}$	**3**
s	$\frac{1}{2}^+$	$-\frac{1}{3}$	$\frac{1}{3}$	**3**	0	0	-1	$-\frac{2}{3}$	**3**

SU(3) (but with quarks whose charges are integral and color dependent) was given in

73. **"Three-triplet Model with Double SU(3) Symmetry," M. Y. Han and Y. Nambu, Phys. Rev. B **139**, 1006–1010 (1965). (A)

74. **"A Systematics of Hadrons in Subnuclear Physics,"** Y. Nambu, in *Preludes in Theoretical Physics*, edited by A. de Shalit, H. Feshbach, and L. van Hove (North-Holland, Amsterdam, 1966), pp. 133–142. (A)

The color terminology was introduced in

75. **Unitary Symmetry and Elementary Particles**, D. B. Lichtenberg (Academic, New York, 1970), pp. 227–228. (I)

but was not generally adopted until the appearance of the articles

76. **"Quarks,"** M. Gell-Mann, in *Elementary Particle Physics*, edited by P. Urban (Springer, Vienna, 1972); Acta Phys. Austriaca Suppl. **9**, 733–761 (1972). (A)

77. **"Light Cone Current Algebra, π^0 Decay and e^+e^- Annihilation,"** W. A. Bardeen, H. Fritzsch, and M. Gell-Mann, in *Scale and Conformal Symmetry in Hadron Physics*, edited by R. Gatto (Wiley, New York, 1973), pp. 139–151. (A)

The quantum-mechanical equivalence of parastatistics of a given order p with an exact internal SU(p) or SO(p) symmetry is reviewed on pp. 84–88, Ref. 78 below.

Color is reviewed in

78. **"Color Models of Hadrons,"** O. W. Greenberg and C. A. Nelson, Phys. Reports **32C**, 69–121 (1977). (A)

as well as in Refs. 163 and 164 below.

III. Quarks as Static Constituents of Hadrons

Quarks have the smallest nonzero spin, $\hbar/2$, small electric charges, $2e/3$ and $-e/3$ (where e is the charge of the proton), small baryon number, 1/3, and the three original flavors of quarks belong to the smallest nontrivial representation of the flavor symmetry group $SU(3)_f$, the triplet representation; thus quarks can be used to construct all hadrons. We give properties of the three original quarks, u (up), d (down), and s (strange), in Table I.

Mesons are constructed as $q\bar{q}$ composites; for example, the pseudoscalar ($J^P = 0^-$) meson octet is given in Table II.

Baryons are constructed as qqq composites; for example, the nucleon $J^P = \frac{1}{2}^+$ octet is given in Table III.

The steps of combining spin as described by $SU(2)_s$ and flavor as described by $SU(3)_f$ in a description using the associative covering group SU(6) of these two groups was taken in

*79. **"Spin and Unitary Spin Independence of Strong Interactions,"** F. Gürsey and L. A. Radicati, Phys. Rev. Lett. **13**, 173–175 (1964). (A)

80. **"Supermultiplets of Elementary Particles,"** B. Sakita, Phys. Rev. B **136**, 1756–1760 (1964). (A)

81. **"Fractionally Charged Particles and SU_6,"** G. Zweig, in *Symmetries in Elementary Particle Physics*, Proc. "Ettore Majorana" International School of Physics, Erice, 1964, edited by A. Zichichi (Academic, New York, 1965), pp. 192–243. (A)

The SU(6) theory and its generalizations are reviewed in

82. **Symmetry Groups in Nuclear and Particle Physics,** F. J. Dyson (Benjamin, New York, 1966). (A)

which contains reprints of important articles. The paradox which led to the introduction of the color degree of freedom for quarks became evident in the context of the SU(6) theory, as discussed above. The introduction of the color degree of freedom, and the associated introduction of the symmetric quark model for baryons, also discussed above, gave impetus to the development of the realistic quark model.

The use of quarks as constituents to describe static properties of hadrons, such as magnetic moments, masses, decay rates, and spectroscopy generally is discussed in detail using phenomenological SU(6) techniques in Refs. 47 and 48 and in

83. **"Quark Models for the Elementary Particles,"** R. H. Dalitz, in *High Energy Physics*, edited by C. DeWitt and M. Jacob (Gordon and Breach, New York, 1965), pp. 253–322. (A)

The first use of the type of SU(6) symmetry-breaking hyperfine interactions which occur in quantum chromodyna-

Table II. Pseudoscalar meson octet as quark composites.

Meson	Quark wave function	I	I_3	Y	S
π^+	$u\bar{d}$	1	1	0	0
π^0	$2^{-1/2}(u\bar{u} - d\bar{d})$	1	0	0	0
π^-	$d\bar{u}$	1	-1	0	0
K^+	$u\bar{s}$	$\frac{1}{2}$	$\frac{1}{2}$	1	1
K^0	$d\bar{s}$	$\frac{1}{2}$	$-\frac{1}{2}$	1	1
$\bar{K}^0$	$s\bar{d}$	$\frac{1}{2}$	$\frac{1}{2}$	-1	-1
K^-	$s\bar{u}$	$\frac{1}{2}$	$-\frac{1}{2}$	-1	-1
η^0	$6^{-1/2}(u\bar{u} + d\bar{d} - 2s\bar{s})$	0	0	0	0

Table III. Nucleon baryon octet as quark composites.

Baryon	Quark composition	I	I_3	Y	S
p	uud	$\frac{1}{2}$	$\frac{1}{2}$	1	0
n	udd	$\frac{1}{2}$	$-\frac{1}{2}$	1	0
Λ^0	uds	0	0	0	-1
Σ^+	uus	1	1	0	-1
Σ^0	uds	1	0	0	-1
Σ^-	dds	1	-1	0	-1
Ξ^0	uss	$\frac{1}{2}$	$\frac{1}{2}$	-1	-2
Ξ^-	dss	$\frac{1}{2}$	$-\frac{1}{2}$	-1	-2

mics to analyze hadron spectroscopy is in a discussion of the ground-state **56** baryon multiplet by

84. **"Hadron Masses in a Gauge Theory,"** A. DeRújula, H. Georgi, and S. L. Glashow, Phys. Rev. D **12**, 147–162 (1975). (A)

Higher Baryon multiplets were discussed from this point of view in a series of articles by N. Isgur and G. Karl reviewed in

85. **"Quarks in Light Baryons," N. Isgur and G. Karl, in *Recent Developments in High-Energy Physics*, Proc. Orbis Scientiae 1980, edited by A. Perlmutter and L. F. Scott (Plenum, New York, 1980), pp. 61–66. (A)

as well as in Ref. 86 below. These articles also go beyond the SU(6) analysis by taking into account effects due to quark mass differences in baryon wave functions as well as in the mass operator, and achieve a remarkably economical and accurate description of the baryon spectrum.

For the ground-state SU(6) multiplet, **56**, in which the quarks are all in s states, only one three-quark state occurs for each allowed value of total angular momentum J isospin I, and hypercharge Y. For higher multiplets, such as the SU(6) **70** with orbital angular momentum $L = 1$, more than one quark model state can have given values of J, I, and Y, and the observed baryons will be superpositions (sometimes called "mixtures") of these quark model states. These superpositions are calculated in mass analyses of the baryon spectrum such as those cited above. In addition to mixtures within an (SU(6),L) multiplet, spin-dependent interactions can produce mixtures between states in different (SU(6),L) multiplets (called "configuration mixing"); these effects are discussed in the review

86. **"Soft QCD: Light Quark Physics with Chromodynamics,"** N. Isgur, University of Toronto preprint (1981), pp 1–21; to appear in Proc. XVI Recontre de Moriond, Les Arcs (1981). (A)

Since mixing effects are very important in the analysis of decays, decay analyses are often coupled with mass spectrum analyses. An analysis of baryon decays is given in

87. **"Baryon Decays in a Quark Model with Chromodynamics,"** R. Koniuk and N. Isgur, Phys. Rev. D **21**, 1868–1886 (1980). (A)

Magnetic moments of baryons were first discussed in the quark model by

88. **"SU(6) and Electromagnetic Interactions,"** M. A. B. Bég, B. W. Lee, and A. Pais, Phys. Rev. Lett. **13**, 514–517 (1964) (Erratum, p. 650). (A)
89. **"Electromagnetic Properties of Baryons in the Supermultiplet Scheme of Elementary Particles,"** B. Sakita, Phys. Rev. Lett. **13**, 643–646 (1964). (A)

These authors independently found the celebrated relation

$$\mu(p)/\mu(n) = -\tfrac{3}{2},$$

between the magnetic moments of the proton and neutron using the SU(6) theory, Refs. 79–81. The concrete interpretation of this result in terms of constituent quarks, and the suggestion that the effective constituent quark mass (for equal mass quarks) is of the order of one-third of the mass of the baryon in which the quarks are bound was given in Ref. 47. A justification of this suggestion on the basis of the Dirac equation for quarks bound in a scalar potential was given by

90. **"Magnetic Moments in Relativistic Quark Models of Elementary Particles,"** H. J. Lipkin and A. Tavkhelidze, Phys. Lett. **17**, 331–332 (1965). (A)

who showed that, for scalar binding, the effective mass which appears in the Dirac magnetic moment is greatly reduced from the mass of a free (unconfined) quark. A discussion of the magnetic moment of a bound state containing a spin-$\frac{1}{2}$ particle for other types of binding (tensor, pseudoscalar, and pseudovector as well as the usual vector case) is given in

91. **"Magnetic Moments of Spin-$\frac{1}{2}$ Particles in External Fields,"** O. W. Greenberg, Phys. Lett. **19**, 423–424 (1965). (A)

Charge radii and form factors of mesons are discussed in

92. **"Electromagnetic Properties of Baryons and Mesons in a Nonrelativistic Quark Model,"** S. B. Gerasimov, Sov. Phys. JETP **23**, 1040–1043 (1966). (A)
93. **"Quark Model Relation for Kaon Charge Radii,"** O. W. Greenberg, S. Nussinov, and J. Sucher, Phys. Lett. **70B**, 465–468 (1977). (A)

Charge radii and form factors of baryons are discussed in

94. **"SU(6)—Strong Breaking Structure Functions and Static Properties of the Nucleon,"** A. Le Yaouanc, L. Oliver, O. Pene, and J. C. Raynal, in *Quarks and Hadronic Structure*, Proc. Erice, 1975, edited by G. Morpurgo (Plenum, New York, 1977), pp. 167–194. (A)
95. **"Why the Neutron Isn't All Neutral,"** R. D. Carlitz, S. D. Ellis and R. Savit, Phys. Lett. **68B**, 443–446 (1977). (A)

Although this section is devoted to static properties of hadrons, we include a reference to the additivity assumption for high-energy hadron–hadron scattering, which states that the elastic scattering amplitude for hadron–hadron scattering is the sum of the elastic scattering amplitudes for the scattering of each quark in one hadron against each quark in the other hadron, weighted by the form factors for each quark in its hadron at the momentum transfer of the given quark. The additivity assumption leads via the optical theorem to relations between asymptotic total cross sections which are in surprisingly good agreement with experiment. Since the actual calculation of the relations at zero momentum transfer to get total cross sections involves counting the number of quarks of each flavor in each hadron, these relations are sometimes called "quark counting" relations. These relations are reviewed in Ref. 24. The spirit of the above calculations is closer to the static quark model than to the parton model which we consider in Sec. IV.

IV. QUARKS AS PARTONS

Quarks as constituents of static hadrons behave as strongly bound massive objects. The complementary view of quarks as "partons" (i.e., "parts of" hadrons) which behave as weakly bound objects, which in some cases can be treated as massless, was introduced in

96. **"The Behavior of Hadron Collisions at Extreme Energies,"** R. P. Feynman, in *High Energy Collisions*, edited by C. N. Yang, J. A. Cole, and M. Good (Gordon and Breach, New York, 1969), pp. 237–258. (A)
97. **"Very High-Energy Collisions of Hadrons,"** R. P. Feynman, Phys. Rev. Lett. **23**, 1415–1417 (1969). (A)

Important early discussions of the quark–parton model include the articles

*98. **"Inelastic Electron–Proton and γ–Proton Scattering and the Structure of the Nucleon,"** J. D. Bjorken and E. A. Paschos, Phys. Rev. **185**, 1975–1982 (1969). (A)
99. **"High Energy Inelastic Neutrino–Nucleon Interactions,"** J. D. Bjorken and E. A. Paschos, Phys. Rev. D **1**, 3151–3160 (1970). (A)
100. **"Inelastic Lepton–Nucleon Scattering and Lepton Pair Production in the Relativistic Quark–Parton Model,"** J. Kuti and V. F. Weisskopf, Phys. Rev. D **4**, 3418–3439 (1971). (A)

and the book

101. **Photon–Hadron Interactions,** R. P. Feynman (Benjamin, Reading, MA, 1972). (A)

The quark–parton model was first introduced to explain Bjorken scaling in electromagnetic and weak inclusive scattering cross sections at high energy. For an exclusive cross section, each particle in the final state is detected. For an inclusive cross section (see Ref. 101), only some of the particles in the final state are detected, so that an inclusive cross section adds up the cross sections to produce the detected particles plus varying numbers of undetected particles. The simplest type of inclusive cross section is a total cross section $\sigma(ab \to X)$ for which none of the particles in the final state is detected and σ depends on the total center-of-mass (CM) energy only. For the inclusive electroproduction cross section $d\sigma(eN \to eX)/d\nu\, dQ^2$ in which only the scattered electron is detected in the final state, the cross section depends on ν, the electron energy loss in the laboratory frame of reference, and $-Q^2$, the square of the electron invariant momentum transfer. These are determined by the incident and scattered electron energies, E and E', and the electron scattering angle θ, all in the laboratory frame Bjorken scaling, which states that $d\sigma(eN \to eX)/d\nu\, dQ^2$ depends only on $x = Q^2/2M\nu$ for large ν and Q^2 at fixed "scaling variable," x, was predicted by

102. **"Asymptotic Sum Rules at Infinite Momentum,"** J. D. Bjorken, Phys. Rev. **179**, 1547–1553 (1969). (A)

The parton model is parametrized by distribution functions $u(x)$, $d(x)$, $s(x)$, $c(x)$, etc., for various flavors of quark–partons in the proton, as well as by antiquark and gluon distribution functions. Here, the fraction of the target hadron linear momentum carried by a given parton is identified, for light quarks, with the scaling variable x defined above. The definitions of these distribution functions lead to sum rules; for example,

$$\int_0^1 dx[u(x) - \bar{u}(x)] = 2 ,$$

which states that the proton has a net number of two u quarks.

The distribution functions for the other available target, the neutron, are expressed in terms of those for the proton. The quark–partons, which interact both weakly and electromagnetically, contribute to scattering cross sections for these processes as well as to the total momentum of the target nucleon. The gluon–partons also carry momentum, but do not directly interact weakly or electromagnetically. They can interact indirectly after converting to quark–antiquark pairs. Comparison of the total momentum sum rule with the formulas for the quark–parton distribution functions in terms of experimentally measured nucleon structure functions gives the surprising result that the gluons carry about half of the proton momentum, as discussed in

103. **"Current-Algebra Sum Rules Suggested by the Parton Model,"** C. H. Llewellyn Smith, Nucl. Phys. B **17**, 277–288 (1970). (A)
104. **"Inelastic Lepton Scattering in Gluon Models,"** C. H. Llewellyn Smith, Phys. Rev. D **4**, 2392–2397 (1971). (A)

These references also discuss experimentally verified relations between the structure functions for electromagnetic and weak processes which hold because the same parton distribution functions occur in the description of both processes.

The spin-$\frac{1}{2}$ nature of quarks leads to a relation in Ref. 49 between two of the otherwise independent scaling structure functions for a given process. Spin-$\frac{1}{2}$ quark–partons, together with left-handed neutrinos and right-handed antineutrinos, and left-handed weak interaction currents for quarks lead to the prediction of different kinematic dependences on $y = \nu/E$ of the inclusive cross sections due to νq and $\bar{\nu}\bar{q}$ scattering (y independent) and to $\bar{\nu}q$ and $\nu\bar{q}$ [proportional to $(1-y)^2$]. This has been verified by experiment to within 10%. The model also predicts, in agreement with experiment, that the total cross sections for neutrino and antineutrino scattering rise linearly with E, and, in rough agreement with experiment, that $(\sigma^{\nu p} + \sigma^{\nu n})/(\sigma^{\bar{\nu} p} + \sigma^{\bar{\nu} n}) = 3$. These relations will be modified at energies large enough to produce significant charm production, and at energies where the momentum transfers carried by the $W^{\pm}$ and Z^0 bosons are no longer negligible compared to their masses.

The first applications of the parton model, to electroproduction and the analogous process of inclusive neutrino scattering, discussed above, were followed by applications to electron–positron annihilation to hadrons, including the ratio of the total hadronic cross section to the muon-pair cross section,

$$R = \frac{\sigma(e^+e^- \to \text{had})}{\sigma(e^+e^- \to \mu^+\mu^-)} \to \tfrac{1}{4} \sum_{\text{spin } 0} Q_i^2 + \sum_{\text{spin } \frac{1}{2}} Q_i^2 \quad s \to \infty ,$$

and the single-hadron inclusive cross section

$$\frac{d\sigma}{d\Omega} \propto \tfrac{1}{2} \sum_{\text{spin } 0} Q_i^2 (1 - \cos\theta) + \sum_{\text{spin } \frac{1}{2}} Q_i^2 (1 + \cos^2\theta).$$

More detailed jet structure derived from QCD is discussed in

105. **"Jets from Quantum Chromodynamics,"** G. Sterman and S. Weinberg, Phys. Rev. Lett. **39**, 1436–1439 (1977). (A)
106. **"Perturbative Quantum Chromodynamics"** E. Reya, Phys. Rep. **69**, 195 (1981). (A)

The qualitative nature of nucleon structure functions is discussed in Ref. 28, p. 246–250. The small-x region of "wee" partons is discussed in Ref. 101, pp. 136–139. SU(6) symmetry predicts that the neutron-to-proton structure function ratio is $\frac{2}{3}$. This prediction is not valid for the $x \to 0$ limit, where wee partons dominate, nor for the $x \to 1$ limit, where large breaking of SU(6) symmetry also occurs, as discussed in Ref. 101, pp. 149–151, and in

107. **"νW_2 at Small ω' and Resonance Form Factors in a Quark Model with Broken SU(6),"** F. E. Close, Phys. Lett. B **43**, 422–426 (1973). (A)
108. **"Approach to Scaling of the Structure Functions of Neutrons and Protons,"** R. Carlitz, Phys. Rev. Lett. **36**, 1001–1004 (1976). (A)

For $x \to 1$, the nucleon electroproduction structure functions are related to the squares of the elastic electromagnetic form factors $G(Q^2)$ of the nucleon according to the relation

$$F_2(x) \propto (1-x)^p, \quad G(Q^2) \propto Q^{-n} \quad \text{with } 2n = p + 1$$

discovered by

109. **"Connection of Elastic Electromagnetic Nucleon Form Factors at Large Q^2 and Deep Inelastic Structure Functions Near Threshold,"** S. D. Drell and T. M. Yan, Phys. Rev. Lett. **24**, 181–186 (1970). (A)
110. **"Phenomenological Model for the Electromagnetic Structure of the Proton,"** G. West, Phys. Rev. Lett. **24**, 1206–1209 (1970). (A)

and discussed from the standpoint of duality by

111. **"Scaling, Duality, and the Behavior of Resonances in Inelastic Electorn–Proton Scattering,"** E. Bloom and F. J. Gilman, Phys. Rev. Lett. **25**, 1140–1143 (1970). (A)

Since the structure functions depend on ν (or $2m\nu$, m the nucleon mass) and Q^2, it is convenient to represent them by contours in a plane whose axes are labelled by these quantities, as done by

112. **"Phenomenological Investigation of Inelastic Electron–Proton Scattering,"** O. Nachtman, Nucl. Phys. B **28**, 283–290 (1971). (A)

The physical region for electroproduction lies in the first quadrant between the lines $Q^2 = 0$ and $2m\nu$. The $x \to 0$ limit corresponds to $Q \to 0$ and is related to real (on-shell) proton Compton scattering. The line $x = 1$ is related to elastic electron–nucleon scattering. Generically, x corresponds to a line from the origin with slope between 0 and 1. Bjorken scaling states that the structure functions become constant as one moves away from the origin on a line of fixed x. Several other scaling variables have been suggested, including the light-cone variable

$$x_{\rm LC} = Q^2/m[\nu^2 + Q^2)^{1/2} + \nu] ,$$

a variable based on duality,

$$x' = Q^2/(2m\nu + m^2) ,$$

and a variable ξ based on operator product analyses of inelastic scattering. References to several scaling variables are given in Ref. 27, p. 364.

The mechanism for lepton-pair production via quark–antiquark annihilation into virtual photons was proposed in

113. **"Massive Lepton-Pair Production in Hadron–Hadron Collisions at High Energies,"** S. D. Drell and T. M. Yan, Phys. Rev. Lett. **25**, 316–320 (1970). (A)
114. **"Partons and Their Applications at High Energies,"** S. D. Drell and T. M. Yan, Ann. Phys. (NY) **66**, 578–623 (1971). (A)

Qualitative predictions which follow from this mechanism include the predictions that meson–nucleon collisions, should give a larger lepton-pair yield than nucleon–nucleon collisions, because mesons contain more antiquark partons than nucleons contain; that the yield from π^- collisions should be greater than from π^+ collisions, because the square of the electric charge of the $\bar{u}$ antiquark parton in the π^- is greater than the square of the charge of the $\bar{d}$ parton in the π^+. In contrast to electron–positron annihilation, where color increases the hadron cross section by a factor of three, in lepton-pair production, color decreases the cross section by a factor of three because the color of the quark and antiquark must match for annihilation to a photon to take place, as discussed in Ref. 78, pp. 107–108.

Good agreement with the Drell–Yan model for muon-pair production by hadrons is reported in

115. **"Experimental Test of the Drell–Yan Model in $p + W \rightarrow \mu^+\mu^- + X$,"** S. R. Smith, S. Childress, P. M. Mockett *et al.*, Phys. Rev. Lett. **46**, 1607–1610 (1981). (A)

The successes of the quark–parton model in weak and electromagnetic interactions with hadrons were followed by applications to hadron–hadron collisions, for example:

116. **"A Quantum Chromodynamic Approach for the Large Transverse Momentum Production of Particles and Jets,"** R. P. Feynman, R. D. Field, and G. C. Fox, Phys. Rev. D **18**, 3320–3343 (1978). (A)
117. **"Perturbative Quantum Chromodynamics and Applications to Large Momentum Transfer Processes,"** R. D. Field, in *Proc. NATO Advanced Study Institute on Quantum Flavordynamics, Quantum Chromodynamics and Unified Theories* (Plenum, New York, 1980), p. 221. (A)
118. **"Hadron Structure and Soft Processes,"** R. C. Hwa, Lectures given at Conference on Partons in Soft Hadronic Processes, Erice, 1981, Oregon preprint OITS-165. (A)

Exclusive processes, including form factors, were treated using quantum chromodynamics by

119. **"Perturbative Quantum Chromodynamics,"** S. J. Brodsky and G. P. Lepage, in *Proc. SLAC Summer Institute, 1979* (SLAC, Stanford, 1979), p. 133. (A)
120. **"Exclusive Processes in Perturbative Quantum Chromodynamics,"** S. J. Brodsky and G. P. Lepage, Phys. Rev. D **22**, 2157–2198 (1980). (A)

The "counting rules" which give the energy dependence of "hard" (i.e., large momentum transfer) scattering processes at fixed angle, and the momentum transfer dependence of electromagnetic form factors, among other things, are reviewed in

121. **'Large Transverse Momentum Processes,"** D. Sivers, S. J. Brodsky, and R. Blankenbecker, Phys. Rep. C **23**, 1–121 (1976). (A)
122. **"Large p_T Physics: Data and the Constituent Models,"** S. D. Ellis and R. Stroynowski, Rev. Mod. Phys. **49**, 753–775 (1977). (A)

Attempts to treat hadron–hadron collisions based strictly on quantum chromodynamics were made in

123. **"Still QCD-ing,"** H. D. Politzer, in *Proc. 8th Hawaii Topical Conference in Particle Physics, 1979* (University of Hawaii, Honolulu, 1979), p. 268. (A)
124. **"Perturbation Theory and the Parton Model in QCD,"** R. K. Ellis, H. Georgi, M. Machacek *et al.*, Nucl. Phys. B **152**, 285–329 (1979). (A)

Restrictions on the possibility of deriving parton model results from QCD were considered in

125. **"Parton Distribution and Decay Functions,"** J. C. Collins and D. E. Soper, Oregon preprint OITS-166 (1981), pp. 1–87. (A)

Quarks appear in elementary-particle and high-energy physics in several guises: as massive constituent quarks in hadron spectroscopy, as light (or massless) quark–partons in high-energy inclusive scattering, and as "current" quarks, the quark fields with associated parameters such as masses that occur, for example, in the Lagrangian of quantum chromodynamics. Although a definitive unified discussion of quarks based on quantum chromodynamics is not yet available, some discussion of the relation between the different guises in which quarks appear is given in Ref. 94 and in

126. **"The Nucleon as a Bound State of Three Quarks and Deep Inelastic Phenomena,"** G. Altarelli, N. Cabibbo, L. Maiani, and L. Petronzio, Nucl. Phys. B **69**, 531–556 (1974). (A)

Current algebra estimates $m_u = 4.3$ MeV/c^2, $m_d = 7.5$ MeV/c^2, and $m_s = 150$ MeV/c^2, of quark masses are given in

127. **"The Problem of Mass,"** S. Weinberg, in *Festschrift for I. I. Rabi*, edited by L. Motz (New York Academy of Science, New York, 1977); Trans. N. Y. Acad. Sci., Ser. II **38**, 185–210 (1977). (A)

These light current quark estimates contrast with the heavier constituent quark masses which follow from hadron mass formulas, or from baryon magnetic moments,

$$m_u \simeq m_d = 336 \text{ MeV}/c^2, \quad m_s = 429\text{–}511 \text{ MeV}/c^2.$$

Relativistic effects which are important if quark masses are light are discussed in Refs. 94 and 126. Reviews of the parton model are given in

*128. **"Evidence for Quarks and Gluons,"** H. L. Anderson, Los Alamos preprint LA-8310-MS (1980), pp. 1–32. (I)
129. **"Neutrino Reactions at Accelerator Energies,"** C. H. Llewellyn Smith, Phys. Rep. C **3**, 261–379 (1972). (A)
130. **"Electron Scattering from Atoms, Nuclei and Nucleons,"** G. B. West, Phys. Rep. C **18**, 263–323 (1975). (A)
131. **"The Parton Model,"** T. M. Yan, Ann. Rev. Nucl. Sci. **26**, 199–238 (1976). (A)
132. **"Parton and Hadron Production in e^+e^- Annihilation,"** S. Wolfram, in Proc. 1980 Moriond Conference. (A)
133. **"Perturbative QCD at High Energies,"** A. H. Mueller, Phys. Rep. **73**, 237–368 (1981). (A)

V. SATURATION, CONFINEMENT, AND COLOR VAN DER WAALS FORCES IN POTENTIAL MODELS

In addition to calculations of bound states of quarks (and antiquarks) with direct application to hadron spectroscopy, there is important work on questions of principle concerning bound states of quarks. When quarks were first suggested, the failure to detect particles with fractional charge was interpreted as implying that quarks are very massive, so that there would not be sufficient energy available in accelerator experiments to produce them, and they would also not occur very often in cosmic rays. The relatively small mass of the observed hadrons was thought to come about from binding effects: for example, this might occur in the potential model if the potential is deep and flat at the origin, so that the quarks could have strong binding together with nonrelativistic motion, as discussed in

134. **"Dynamical Symmetries and Fundamental Fields,"** Y. Nambu, in *Symmetry Principles at High Energy*, Proc. 2nd Coral Gables Conference, edited by B. Kursunoglu, A. Perlmutter, and I. Sakmar (Freeman, London, 1967), pp. 274–285. (A)

and Ref. 25. This point of view encounters serious difficulties, including the problem that meson exchange interactions give a Coulomb or Yukawa potential, instead of a potential flat at the origin,

135. **"Non-Relativistic Motion of Particles in Strongly Bound *S*-States,"** O. W. Greenberg, Phys. Rev. **147**, 1077–1080 (1966). (A)
136. **"Quark Models Old and New,"** R. H. Dalitz, in *Fundamentals of Quark Models*, edited by I. M. Barbour and A. T. Davies (Scottish Univ. Summer School Physics, Edinburgh, 1977), pp. 151–244. (A)

Dalitz, in this last article, also points out that the joint requirements of nonrelativistic motion and strong binding do not, as is usually assumed, lead to a normal Schrödinger equation, rather a fully relativistic equation must be used, such as the Bethe–Salpeter equation, which is used in

137. **"Dynamics of the Meson Spectrum,"** H. Joos, in *Quarks and Hadronic Structure*, edited by G. Morpurgo (Plenum, New York, 1977), pp. 203–224. (A)

Nambu discussed color gluon exchange as a mechanism to account for "saturation," the fact that only color-singlet hadrons occur at low mass, in the context of models in which free quarks could, in principle, occur, without taking into account the spatial dependence of the interquark interaction, in Refs. 73, 74, and 134. The central results of Nambu's analysis are that quarks which are heavy in free space can have small effective masses inside hadrons, that the only light-mass hadrons will be color-singlets, and that color-singlet systems will be neutral or inert, i.e., will not experience the strong color interaction mediated by color-octet gluons. Nambu's ideas were explored further by

138. **"Saturation in Triplet Models of Hadrons,"** O. W. Greenberg and D. Zwanziger, Phys. Rev. **150**, 1177–1180 (1966). (A)
139. **"Triality, Exotics and the Dynamical Basis of the Quark Model,"** H. J. Lipkin, Phys. Lett. **45B** 267–271 (1973). (A)
140. **"On Quark Atoms, Molecules and Crystals,"** H. J. Lipkin, Phys. Lett. 58B, 97–99 (1975). (A)
141. **"A Quasinuclear Colored Quark Model for Hadrons,"** H. J. Lipkin, in *Common Problems in Low- and Medium-Energy Nuclear Physics, Proc. NATO Advanced Study Institute, Banff, 1978*, edited by B. Castel, B. Gouland, and F. C. Khanna (Plenum, New York, 1979), p. 173–212. (A)

The introduction of the idea that quarks are permanently confined in hadrons leads to the consideration of potentials which rise indefinitely, and requires important changes in the discussion of saturation. Although the leading large-distance term in the interquark potential is absent in the effective potential between color-singlet hadrons, inverse-power color analog van der Waals forces remain, and these are in strong contradiction with experiment, as discussed and reviewed in

142. **"van der Waals-like Forces Between Hadrons Induced by Color Confining Potentials,"** M. B. Gavela, A. Le Yaouanc, L. Oliver, O. Pène, J. C. Raynal, and S. Sood, Phys. Lett. **82B**, 431–435 (1979). (A)
143. **"Is There a Strong van der Waals Force Between Hadrons?"** G. Feinberg and J. Sucher, Phys. Rev. **20**, 1717–1735 (1979). (A)
144. **"The Potential Model of Colored Quarks: Success for Single-Hadron States, Failure for Hadron–Hadron Interactions,"** O. W. Greenberg and H. J. Lipkin, Nucl. Phys. A **370**, 349–364 (1981). (A)

These articles also discuss pathological behavior which can occur in a system of several quarks (and antiquarks) when the quark–quark potential grows too rapidly (more rapidly than r^2); further discussion of this is in

145. **"Color van der Waals Forces?"** O. W. Greenberg, in *Recent Developments in High-Energy Physics*, edited by A. Perlmutter and L. F. Scott (Plenum, New York, 1980), pp. 67–86. (A)
146. **"Quark Confinement and Hadron Separation,"** O. W. Greenberg and J. Hietarinta, University of Maryland preprint 81-122 (1981), to appear in Proc. 17th Winter School of Theoretical Physics at Karpacz (to be published by Gordon and Breach). (A)

VI. QUANTUM CHROMODYNAMICS AND PERMANENT QUARK AND COLOR CONFINEMENT

The suggestion that quarks might be permanently confined was motivated on the experimental side by the failure to observe isolated quarks. On the theoretical side, it followed the discovery of asymptotic freedom. Asymptotic freedom holds for a field theory if the "running" coupling constant, which depends on the distance or the momentum, vanishes in the limit of arbitrarily small distances or arbitrarily large momenta. Asymptotic freedom was first pointed out by 't Hooft at the 1972 Marseille Conference, but his work was unpublicized and remained unpublished. The result was discovered and published in

147. **"Reliable Perturbative Results for Strong Interactions?"** H. D. Politzer, Phys. Rev. Lett. **30**, 1346–1349 (1973). (A)
148. **"Ultraviolet Behavior of Non-Abelian Gauge Theories,"** D. Gross and F. Wilczek, Phys. Rev. Lett. **30**, 1343–1346 (1973). (A)

and is reviewed in

149. **"Asymptotic Freedom: An Approach to Strong Interactions,"** H. D. Politzer, Phys. Rep. **C14**, 129–180 (1974). (A)

The fact that the asymptotic freedom occurs only in non-Abelian gauge theories was pointed out in

150. **"The Price of Asymptotic Freedom,"** S. Coleman and D. Gross, Phys. Rev. Lett. **31**, 851–854 (1973). (A)

Non-Abelian gauge theories were invented in

151. **"Conservation of Isotopic Spin and Isotopic Gauge Invariance,"** C. N. Yang and R. L. Mills, Phys. Rev. **96**, 191–195 (1954). (A)
152. R. Shaw, Ph.D. Thesis, Cambridge University, 1955 (unpublished). (A)

They are discussed in

153. **"Gauge Theories,"** E. S. Abers and B. W. Lee, Phys. Rep. C **9**, 1–141 (1973). (A)
154. **"Secret Symmetry: An Introduction to Spontaneous Symmetry Breakdown and Gauge Fields,"** S. Coleman, in *Laws of Hadronic Matter*, Proc. 1973 International School of Subnuclear Physics at Erice, edited by A. Zichichi (Academic, New York, 1975), pp. 138–223. (A)
155. **Gauge Theories of Weak Interactions,** J. C. Taylor (Cambridge University, Cambridge, 1976). (A)
156. **Quantum Theory of Gauge Fields,** L. D. Faddeev and A. A. Slavnov (Addison-Wesley, Reading, MA, 1980). (A)
157. **Quantum Field Theory,** C. Itzykson and J.-B. Zuber (McGraw-Hill, New York, 1980). (A)
158. **Field Theory: A Modern Primer,** P. Ramond (Addison-Wesley, Reading, MA, 1981). (A)
159. **Field Theory and Particle Physics,** T. D. Lee (Harwood, New York, 1981). (A)

Since the "running" coupling constant vanishes for small distances in an asymptotically free theory, it seemed plausible to conjecture that the coupling constant could grow without bound for large distances, in particular that severe infrared divergences associated with the massless gluons in exact $SU(3)_c$ gauge theory would lead to a superstrong force between objects carrying color at low momentum transfers, or equivalently, at large distances, and would prevent separation of particles carrying color charges, i.e., would produce permanent confinement. This "infrared slavery" conjecture was put forward in

160. **"Non-Abelian Gauge Theories of the Strong Interactions,"** S. Weinberg, Phys. Rev. Lett. **31,** 494–497 (1973). (A)

161. **"Asymptotically Free Gauge Theories. I",** D. J. Gross and F. Wilczek, Phys. Rev. D **8,** 3633–3652 (1973). (A)

162. **"Advantages of the Color Octet Picture,"** H. Fritzsch, M. Gell-Mann, and H. Leutwyler, Phys. Lett. B **47,** 365–368 (1973).

Elementary discussions of confinement are given in

•163. **"The Confinement of Quarks,"** Y. Nambu, Sci.Am. **235** (5), 48–71 (1976). (E)

•164. **"Quark Confinement,"** R. Jaffe, Nature **268,** 201–208 (1977). (E)

Although the consensus of workers in the field is that confinement is a property of quantum chromodynamics, it has proved difficult to demonstrate this property. Confinement does not emerge in the context of perturbation theory as shown in

165. **"Kinoshita's Theorem in Yang–Mills Theories,"** G. Sterman, Phys. Rev. D **14,** 2123–2125 (1976). (A)

Confinement also does not appear in classical solutions of the Yang–Mills equations with quark and antiquark sources. On the contrary, there is a solution for which the gauge potential behaves as $1/r$ like the Coulomb solution of an Abelian theory, as shown by

166. **"Some Solutions of the Classical Isotopic Gauge Field Equations,"** T. T. Wu and C. N. Yang, in *Properties of Matter Under Unusual Conditions*, edited by H. Mark and S. Fernbach (Interscience, New York, 1968), pp. 349–354. (A)

Surprisingly, this solution becomes unstable above a critical value of the coupling constant, as first shown by

167. **"Color Screening by a Yang–Mills Instability,"** J. E. Mandula, Phys. Lett. **67B,** 175–178 (1977). (A)

The literature on these instabilities can be traced from

168. **"Classical Chromodynamics of Two Sources: Charges and Magnetic Moments,"** R. A. Freedman, L. Wilets, S. D. Ellis, and E. M. Henley, Phys. Rev. D **22,** 3128–3134 (1980). (A)

A review of classical solutions of Yang–Mills theories which discusses monopoles, instantons, and merons, but not the instabilities just mentioned is

169. **"Classical Solutions of SU(2) Yang–Mills Theories,"** A. Actor, Rev. Mod. Phys. **51,** 461–525 (1979). (A)

These failures to find the confining potential are in contrast with the situation in quantum electrodynamics, where the Coulomb potential occurs in the solution of classical electrodynamics as well as in the perturbation expansion of the quantized theory.

The existence of Regge trajectories of resonances for which J, the angular momentum of the resonance grows, in good approximation, linearly with s, the square of the center-of-mass energy of the resonance, provides experimental evidence for confinement, since in the absence of confinement there would be an ionization energy, above which there would no longer be hadronic resonances. Rather, above the ionization energy, the quarks and antiquarks would be unbound and could be observed singly. Linear Regge trajectories occur in dual models, discussed in

•170. **"Dual Resonance Models of Elementary Particles,"** J. H. Schwarz, Sci. Am. 232 (2), 61–69 (February 1975). (E)

171. **Dual Theory,** edited by M. Jacob (North-Holland, Amsterdam, 1974). (A)

which contains reprints of review articles on dual models. The dual models can be interpreted in terms of strings. A number of authors have related the string model to gauge field theory. The suggestion that the hadronic strings correspond to Landau–Ginsburg vortices of quantized magnetic flux in the vacuum of a spontaneously broken Abelian gauge field theory in the semiclassical approximation was given in

172. **"Vortex-Line Models for Dual Strings,"** H. B. Nielsen and P. Olesen, Nucl. Phys. B **61,** 45–61 (1973). (A)

The suggestion of attaching monopoles to finite Nielsen–Olesen vortices is in

173. **"Strings, Monopoles and Gauge Fields,"** Y. Nambu, Phys. Rev. D **10,** 4262–4281 (1974). (A)

In the Nielsen–Olesen model, the vacuum acts as a superconductor, and forces magnetic flux into flux tubes, which are the "strings" of the model. The flux tubes have a constant energy per unit length, and thus the potential between quarks and antiquarks is proportional to the distance between the particles. The flux can be absorbed by monopoles at the ends producing a meson of finite energy. A single monopole would have to have its flux go to infinity along a flux tube, and thus would have infinite energy. If monopoles are identified with quarks, then a single quark would have infinite energy and would not be observed. This picture thus provides a qualitative understanding of confinement.

In quantum chromodynamics, quarks are not monopoles, but rather carry electric color charge. The suggestion that electric and magnetic quantities should be interchanged, so that electric flux would be forced into flux tubes, and these flux tubes should start and end on electric color charges, to be identified with quarks, for states of finite energy, was made in

174. **"Gauge Fields with Unified Weak, Electromagnetic, and Strong Interactions,"** G. 't Hooft, in *High Energy Physics*, Proc. European Phys. Soc. Int. Conf., edited by A. Zichichi (Editrice Compositore, Bologna, 1976), p. 1225. (A)

and developed in

175. **"Vortices and Quark Confinement in Non-Abelian Gauge Theories,"** S. Mandelstam, Phys. Rep. C **23,** 245–249 (1976). (A)

An important criterion for confinement in pure Yang–Mills theories (i.e., without quarks or antiquarks) is that the "Wilson loop"

$$\operatorname{tr} P\left(\exp ig \int_C dx^\mu \lambda^a A_\mu^a(x)\right),$$

where P stands for path ordering, the λ^a are the Gell-Mann SU(3) matrices, and the $A_\mu^a(x)$ are the octet of SU(3) gauge fields, which creates a loop of electric color flux along the curve C around which the integral is evaluated should depend asymptotically on the area enclosed by the loop, rather than on the perimeter of the loop, as would be the case in theories without confinement. The area law was proposed in

176. **"Confinement of Quarks,"** K. G. Wilson, Phys. Rev. D **10,** 2445–2459 (1974). (A)

This article also put forward the lattice gauge theory approach to quantum chromodynamics, and demonstrated confinement in the strong coupling limit in lattice gauge theory. Starting with the article of Wilson, Ref. 176, there has been intensive work on confinement in the context of lattice gauge theory. This work, and lattice gauge theory generally is reviewed in

177. **"Quantum Chromodynamics on a Lattice,"** K. G. Wilson, in *New Developments in Quantum Field Theory and Statistical Mechanics*, Proc. NATO Advanced Study Institute, Cargèse, 1976, edited by M. Levy and P. Mitter (Plenum, New York, 1977), pp. 143–172. (A)

178. **"The Application of Renormalization Group Techniques to Quarks and Strings,"** L. P. Kadanoff, Rev. Mod. Phys. **49,** 267–296 (1977). (A)

179. **"An Introduction to Lattice Gauge Theory and Spin Systems,"** J. B. Kogut, Rev. Mod. Phys. **51,** 659–714 (1979). (A)

't Hooft suggested consideration of an operator which creates a tube of magnetic flux, and derived commutation relations between his operator and Wilson's operator. 't Hooft used these commutation relations to classify the possible area and/or perimeter behaviors of the operators, and related these behaviors to the confinement and other possible phases of the theory. This work is reviewed in

180. **"The Topological Mechanism for Permanent Quark Confinement in Quantum Chromodynamics,"** G. 't Hooft, Utrecht preprint, 1980, pp. 1–58, lectures at 21st Scottish Universities Summer School in Physics, St. Andrews, 1980 (unpublished). (A)

Quantum chromodynamics can be formulated in any dimension of space-time, and for any SU(N) or U(N) group. Planar graphs without quark loops dominate the limit in which $N\rightarrow\infty$, and g^2N = const, and thus in this limit quantum chromodynamics simplifies, as shown in

181. **"A Planar Diagram Theory for Strong Interactions,"** G. 't Hooft, Nucl. Phys. B **72,** 461–473 (1974). (A)

and discussed further in

182. **"Baryons in the $1/N$ Expansion,"** E. Witten, Nucl. Phys. B **160,** 57–115 (1979). (A)

183. **"$1/N$,"** S. Coleman, SLAC-PUB-2484, 1980, pp. 1–70; to appear in Proc. 1979 International School of Subnuclear Physics, Erice. (A)

In two space-time dimensions, the gluon fields do not carry dynamical degrees of freedom, and quantum chromodynamics in the large-N limit has been solved exactly, in

184. **"A Two-Dimensional Model for Mesons,"** G. 't Hooft, Nucl. Phys. B **75,** 461–470 (1974). (A)

The linear potential is the one space dimension analog of the Coulomb potential, and thus confinement is automatic in this case. An attempt to carry over the mechanism of one-dimensional quantum chromodynamics to three-dimensional space was made in

185. **"Vacuum Polarization and the Absence of Free Quarks in Four Dimensions,"** J. Kogut and L. Susskind, Phys. Rev. D **9,** 3501–3512 (1974). (A)

A program of work based on the idea that instantons play a central role in producing confinement was carried out in

186. **"Instantons as a Bridge Between Weak and Strong Coupling in Quantum Chromodynamics,"** C. G. Callan, R. F. Dashen, and D. J. Gross, Phys. Rev. D **20,** 3279–3289 (1979). (A)

Instantons are discussed in

*187. **"Solitons,"** C. Rebbi, Sci. Am. **240** (2), 92–116 (1979). (E)

188. **"The Uses of Instantons,"** S. Coleman, in *The Whys of Nuclear Physics*, Proc. 1977 International School of Subnuclear Physics, edited by A. Zichichi (Plenum, New York, 1979), p. 805. (A)

189. **"Instantons and Monopoles in Yang–Mills Gauge Field Theories,"** M. K. Prasad, Physica D **1,** 167–191 (1980). (A)

Monte Carlo calculations of lattice quantum chromodynamics have given numerical evidence for confinement, as described in

190. **"Monte Carlo Study of Quantized SU(2) Gauge Theory,"** M. Creutz, Phys. Rev. D **21,** 2308–2315 (1980). (A)

191. **"Numerical Techniques for Lattice Gauge Theories,"** M. Creutz, BNL-29048, 1981, pp. 1–44; lectures given at 18th Winter School of Theoretical Physics, Karpacz, 1981 (unpublished). (A)

A general review of quantum chromodynamics approaches to confinement is given in

*192. **"General Introduction to Confinement,"** S. Mandelstam, Phys. Rep. **67,** 109–121 (1980). (A)

Physical effects of gluons, the color-carrying mediators of the strong force in quantum chromodynamics, are discussed in

193. **"Gluons,"** J. Ellis, TH. 2817, 1980, pp. 1–20 (to appear in Comments on Nucl. and Part. Phys.). (A)

The status of experimental searches for glueballs, color-singlet particles which are composed only of gluons, is summarized in

194. **"Glueballs, A Little Review,"** P. M. Fishbane, University of Virginia preprint, 1981, pp. 1–26 (to appear in Proc. 1981 Orbis Scientiae, University of Miami at Coral Gables). (A)

Reviews of quantum chromodynamics are given in Refs. 163 and 164, and in

195. **"Hard Processes in Quantum Chromodynamics,"** Yu. L. Dokshitzer, D. I. Dyakonov, and S. I. Troyan, Phys. Rep. **58,** 269–395 (1980). (A)

196. **"Quantum Chromodynamics and Its Applications,"** J. Ellis and C. T. Sachradja, in *Quarks and Leptons*, Proc. 1979 Cargèse Summer Inst., edited by M. Lévy, J.-L. Basdevant, D. Speiser, and J. Weyers (Plenum, New York, 1980), pp. 285–432. (A)

197. **"Quantum Chromodynamics,"** W. Marciano and H. Pagels, Phys. Rep. C **36,** 137–276 (1978). (A)

198. **"Perturbative QCD,"** C. H. Llewellyn Smith, in *High Energy Physics 1980*, Proc. XX International Conference, Madison, 1980, edited by L. Durand and L. G. Pondrum (American Institute of Physics, New York, 1981), pp. 1345–1377. (A)

199. **"Asymptotic Freedom in Deep Inelastic Processes in the Leading Order and Beyond,"** A. J. Buras, Rev. Mod. Phys. **52,** 199–276 (1980). (A)

Because the demonstration that QCD leads to permanent confinement of quarks (and objects with nonsinglet color generally) has been so difficult, phenomenological approaches to confinement have been pursued. Prominent among these approaches is the bag model, which has been developed in different versions. In bag models, qualitative properties of confinement, which are assumed to follow from QCD, are introduced in an *ad hoc* manner. For example, in the MIT bag model, hadrons are bubbles in space that contain quarks and antiquarks in color-singlet states; inside the bubbles, the quarks interact relatively weakly via color–gluon exchange. Nonsinglet combinations of quarks are absolutely prohibited from leaving a bag. Bag models are reviewed in

*200. **"The Bag Model of Quark Confinement,"** K. A. Johnson, Sci. Am. **241** (1), 112–121 (1979). (E)

201. **"A Practical Model of Quark Confinement,"** R. L. Jaffe and K. A. Johnson, Comm. Nucl. Part. Phys. **7,** 107–119 (1977). (I)

202. **"The Quark Bag Model,"** P. Hasenfratz and J. Kuti, Phys. Rep. C **40,** 75–179 (1978). (A)

203. **"Applications of the Bag Model,"** R. L. Jaffe, MIT preprint MIT-CTP-846, 1980, pp. 1–76 (to appear in Proc. 19th Schladming Winter School, 1980). (A)

VII. UNIFIED THEORIES

The first step in giving a unified theoretical description of different interactions was the partial unification of the electromagnetic and weak interactions, the $SU(2)_L \times U(1)$ electroweak theory, which was developed in

204. **"Partial-Symmetries of Weak Interactions,"** S. L. Glashow, Nucl. Phys. **22,** 579–588 (1961). (A)

205. **"A Model of Leptons,"** S. Weinberg, Phys. Rev. Lett. **19,** 1264–1266 (1967). (A)

206. **"Weak and Electromagnetic Interactions,"** A. Salam, in *Elementary Particle Theory, Proc. of the Eighth Nobel Symposium*, edited by N. Svartholm (Almquist and Wiksell, Stockholm, 1968), pp. 367–377. (A)

The renormalizability of spontaneously broken (i.e., broken via the Higgs–Kibble mechanism) Yang–Mills theories was first pointed out by

207. **"Renormalizable Lagrangian for Massive Yang–Mills Fields,"** G. 't Hooft, Nucl. Phys. B **35**, 167–188 (1971). (A)

Further discussion and review of renormalization for such theories is given in Ref. 214 below. There is now a large literature on the electroweak theory, which is part of the "standard model." An elementary introduction is given in

*208. **"Unified Theories of Elementary Particle Interactions,"** S. Weinberg, Sci. Am. **231** (1), 50–59 (1974). (E)

Good reviews appear in the 1979 Nobel lectures of the above authors,

209. **"Conceptual Foundations of the Unitied Theory of Weak and Electromagnetic Interactions,"** S. Weinberg, Rev. Mod. Phys. **52**, 515–523 (1980). (A)

210. **"Gauge Unification of Fundamental Forces,"** A. Salam, Rev. Mod. Phys. **52**, 525–538 (1980). (A)

211. **"Towards a Unified Theory: Threads in a Tapestry,"** S. L. Glashow, Rev. Mod. Phys. **52**, 539–543 (1980). (A)

as well as in

*212. **"The 1979 Nobel Prize in Physics,"** S. Coleman, Science **206**, 1290–1292 (1979). (I)

213. **"Spontaneous Symmetry Breaking, Gauge Theories, the Higgs Mechanism, and All That,"** J. Bernstein, Rev. Mod. Phys. **46**, 7–48 (1974). (I)

214. **"Gauge Theories,"** E. S. Abers and B. W. Lee, Phys. Rep. C **9**, 1–141 (1973). (A)

215. **Gauge Theories of Weak Interactions,** J. C. Taylor (Cambridge University, Cambridge, 1976). (A)

216. **"Introduction to Gauge Theories of Electromagnetic and Weak Interactions,"** L.-F. Li, University of California at Santa Barbara Preprint NSF-ITP-81-25, 1981, pp. 1–188; lectures given at the 1981 Summer School on Particle Physics, Hefei, China (preliminary version). (A)

A review which emphasizes the possibility that the experimental verification of the low-energy predictions of the electroweak model need not imply the validity of the full content of the model is given in

217. **"Electroweak Interactions,"** J. D. Bjorken, FERMILAB-Conf-80/86-THY, 1980 (unpublished). (A)

The idea that the symmetry breaking of the $SU(2)_L \times U(1)$ unified theory of weak and electromagnetic interactions occurs through a new "hypercolor" or "technicolor" interaction was suggested in

218. **"Implications of Dynamical Symmetry Breaking,"** S. Weinberg, Phys. Rev. D **13**, 974–996 (1976). (A)

219. **"Implications of Dynamical Symmetry Breaking: An Addendum,"** S. Weinberg, Phys. Rev. D **19**, 1277–1280 (1978). (A)

220. **"Dynamics of Spontaneous Symmetry Breaking in the Weinberg-Salam Theory,"** L. Susskind, Phy. Rev. D **20**, 2619–2625 (1979). (A)

Although considerable work has been done on this idea, most workers do not think a realistic model of the hypercolor mechanism has been found. Reviews of this work are given in

221. **"An Introduction to Technicolour,"** P. Sikivie, TH. 2951-CERN, 1980, pp. 1–41 (to appear in Proc. International School of Physics "Enrico Fermi", Varenna, 1980). (A)

222. **"An Introduction to Weak Interaction Theories with Dynamical Symmetry Breaking,"** K. D. Lane and M. E. Peskin, NORDITA preprint NORDITA-80/33, 1980, pp. 1–46 (to appear in Proc. XVth Recontre de Moriond, 1980). (A)

The next step was to include the strong interactions, via the Yang–Mills gauge theory for the color symmetry SU(3), in the unification program. This was first done using the local gauge symmetry group $SU(4)\times SU(4)$, in which charm is added to the original three quark flavors and lepton number is added to the original three colors, with the quarks and leptons unified in the fundamental matter multiplet (4,4) in

223. **"Unified Lepton-Hadron Symmetry and a Gauge Theory of the Basic Interactions,"** J. C. Pati and A. Salam, Phys. Rev. D **8**, 1240–1251 (1973). (A)

224. **"Lepton Number as the Fourth 'Color',"** J. C. Pati and A. Salam, Phys. Rev. D **10**, 275–289 (1974). (A)

The latter paper is remarkable in containing the first suggestion that quarks and leptons may be composites of the *same* set of more fundamental fields, and in suggesting why the weak interactions are chiral, while the strong interactions are vector, in Footnote 7, p. 287.

Another path towards the unification of the fundamental interactions was taken by searching for a minimal unifying gauge group which contains $SU(3)_C \times SU(2)_L \times U(1)$. The SU(5) theory, which gives a very economical scheme of unification, was proposed in

225. **"Unity of All Elementary Particle Forces," H. Georgi and S. L. Glashow, Phys. Rev. Lett. **32**, 438–441 (1974). (A)

The important step of showing that all three of the coupling constants of the strong, electromagnetic, and weak interactions, associated with the $SU(3)_C$,$SU(2)_L$, and $U(1)_{em}$ gauge interactions, respectively, which depend on mass via the renormalization group, assume equal values at a single mass (the "unification mass") in theories in which the symmetry breaking occurs in one step (and certain other conditions hold) was taken in

*226. **"Hierarchy of Interactions in Unified Gauge Theories,"** H. Georgi, H. Quinn, and S. Weinberg, Phys. Rev. Lett. **33**, 451–454 (1974). (A)

The fact that the unification mass is of order 10^{16} GeV, while the lower limit of order 10^{30} years for the proton lifetime requires a vector boson mass greater than 10^{15} GeV is an encouraging sign for theories of the type just mentioned. The value of the Weinberg angle θ_W (see Refs. 213–216 for the definition of θ_W) at low energies can be calculated in a given grand unified theory using the structure of the theory at the unification mass, together with the renormalization group equations. This was first done in Ref. 226. A detailed analysis of $\sin^2 \theta_W$ is given in

227. **"Precise SU(5) Predictions for $\sin^2 \theta_W, m_W$, and m_Z,"** W. Marciano and A. Sirlin, Phys. Rev. Lett. **46**, 163–166 (1981). (A)

Other unified models were proposed, using the gauge group SO(10), which allows the quarks and leptons of a given generation to be placed in a single irreducible representation (together with a right-handed neutrino), and thus allows massive neutrinos. The SO(10) model also allows a left–right symmetric model of weak and electromagnetic interactions, like the $SU(4)\times SU(4)$ model of Refs. 223 and 224. The SO(10) model is discussed in

228. **"The State of the Art—Gauge Theories,"** H. Georgi, in *Particles and Fields*, edited by C. E. Carlson (American Institute of Physics, New York, 1975), pp. 575–582. (A)

One of the most striking predictions of the "grand unified" theories, i.e., theories which unify the strong, electromagnetic, and weak interactions in a gauge theory which has only one coupling constant and in which quarks and leptons are in the same multiplet(s), is the prediction of proton decay. Proton decay was first discussed in this context in Ref. 223, and was discussed further from a somewhat different point of view in

*229. **"Is Baryon Number Conserved?,"** J. C. Pati and A. Salam, Phys. Rev. Lett. **31**, 661–664 (1973). (A)

Proton decay had been discussed, but not in the context of unified gauge theories, together with an analysis of the conditions under which a baryon asymmetry can arise from a universe which is initially baryon–antibaryon symmetric, in the remarkable article

*230. **"Violation of *CP*-Invariance, *C* Asymmetry, and Baryon Asymmetry of the Universe,"** A. D. Sakharov, JETP Lett. **5**, 24–27 (1967). (A)

An elementary introduction to proton decay is given in

*231. **"The Decay of the Proton,"** S. Weinberg, Sci. Am. **244** (6), 64–75 (1981). (E)

A detailed discussion of the proton lifetime in the SU(5) and related models appears in

232. **"Uncertainties in the Proton Lifetime,"** J. Ellis, M. K. Gaillard, D. V. Nanopoulos, and S. Rudaz, Nucl. Phys. B **176,** 61–99 (1980). (A)

The different possibilities for baryon (and lepton) nonconserving processes are discussed in

233. **"Baryon and Lepton Nonconserving Processes,"** S. Weinberg, Phys. Rev. Lett. **43,** 1566–1570 (1979). (A)

234. **"Operator Analysis of Nucleon Decay,"** F. Wilczek and A. Zee, Phys. Rev. Lett. **43,** 1571–1573 (1979). (A)

235. **"Varieties of Baryon and Lepton Nonconservation,"** S. Weinberg, Phys. Rev. D **22,** 1694–1700 (1980). (A)

A general discussion of processes which violate baryon number and/or lepton number conservation, including, for example, neutron–antineutron oscillations, is given in

236. **"Probing the Design of Grand Unification Through Conservation Laws,"** J. C. Pati, in *Weak Interactions as Probes of Unification*, Proc. Virginia Polytechnic Institute Conference, 1981, edited by G. B. Collins, L. N. Chang, and J. R. Ficenec (American Institute of Physics, New York, 1981), pp. 84–106. (A)

An outstanding puzzle is the generation or family structure of quarks and leptons. Two quark generations, and half of a third generation have been observed; and three lepton generations have been observed. Each generation has the same $SU(3)_{color} \times SU(2)_L \times U(1)$ quantum numbers. The generations are $(u,d;\nu_e,e)$, $(c,s;\nu_\mu,\mu)$, $(t,b;\nu_\tau,\tau)$, where the first two entries are the quarks, and the last two entries are the leptons, and the t (top) quark has not been observed. An analysis of neutral generation-changing processes is given in

237. **"Bounds on the Masses of Neutral Generation-Changing Gauge Bosons,"** R. N. Cahn and H. Harari, Nucl. Phys. B **176,** 135–152 (1980). (A)

Elementary descriptions of grand unified theories are given in

*238. **"Gauge Theories of the Forces between Elementary Particles,"** G. 't Hooft, Sci. Am. **242** (6), 104–138 (1980). (E)

*239. **"A Unified Theory of Elementary Particles and Forces,"** H. Georgi, Sci. Am. **244** (1), 48–63 (1981). (E)

A systematic study of grand unified theories and their implications for proton decay is given in

240. **"Color Embeddings, Charge Assignments, and Proton Stability in Unified Gauge Theories,"** M. Gell-Mann, P. Ramond, and R. Slansky, Rev. Mod. Phys. **50,** 721–744 (1978). (A)

Among other reviews of unified theories are

241. **"Grand Unified Theories and Proton Decay,"** P. Langacker, Phys. Rep. **72,** 185–385 (1981). (A)

242. **"Grand Unified Theories,"** J. Ellis, CERN TH.-2942, 1980, pp. 1–102 (to appear in Proc. 21st Scottish Universities Summer School in Physics). (A)

243. **"Introduction to Gauge Theories of the Strong, Weak, and Electromagnetic Interactions,"** C. Quigg, FERMILAB-Conf-80/64-THY, 1980, pp. 1–136 [to be published in *Proc. NATO Advanced Studies Institute on Techniques and Concepts of High Energy Physics*, St. Croix, 1980, edited by T. Ferbel (Plenum, New York, 1981)]. (A)

Supersymmetry, a type of symmetry which transforms between Bose and Fermi fields, might play a role in elementary particle physics.

Supersymmetry is reviewed in

244. **"Supersymmetry and Superfields,"** A. Salam and J. Strathdee, Fortsch. Phys. **26,** 57 (1978). (A)

245. **"Supersymmetry,"** P. Fayet and S. Ferrara, Phys. Rep. C **32,** 249–334 (1977). (A)

The fact that the unification mass is of order 10^{16} GeV, while the Planck mass, the characteristic mass of quantum gravity, is of order 10^{19} GeV, raises the possibility that gravity should be unified with the other interactions. This possibility is considered in

246. **"Supergravity and the Unification of the Laws of Physics,"** D. Z. Freedman and P. van Nieuwenhuizen, Sci. Am. **238** (2), 126–143 (1978). (A)

247. **"Attempts at Superunification,"** J. Ellis, M. K. Gaillard, L. Maiani, and B. Zumino, in *Unification of the Fundamental Particle Interactions (1980)*, edited by S. Ferrara, J. Ellis, and P. Van Nieuwenhuizen (Plenum, New York, 1980), pp. 69–88. (A)

VIII. NEXT LAYER OF THE ONION?: POSSIBLE QUARK COMPOSITENESS

The repetition of families of quarks and leptons is not explained convincingly by the unified models presently available. In addition, even at the quark–lepton level, there are a large number of states. Counting a color-triplet quark of a given flavor as three states, a single family of quarks and leptons contains 16 left-handed states (15, if the neutrino is massless); thus three generations contain 48 left-handed states (45, if the neutrinos are massless). The photon, eight colored gluons, $W^{\pm}$ and Z^0 vector bosons account for at least 12 more vector particles, each of which has two (the photon and color gluons) or three (the $W^{\pm}$ and Z^0 bosons) helicity states. Other vector bosons which occur in unified models, as well as Higgs bosons, add still more states to this list. Composite models of quarks, leptons, and other particles might allow new understanding both of the repetition of families, and of the large number of states. Such models also provide a new point of view concerning unification: hadronic and leptonic matter can be composed of a common set of constituents.

Early articles on composite quarks and leptons include Ref. 224 above, which noted the possibility of expressing their unified model in terms of subconstituent fields in their Footnote 7. They chose a spin-$\frac{1}{2}$ quartet to carry flavor and a spin-0 quartet to carry color (including lepton number). They introduced the idea that the same fields can carry flavor for both quarks and leptons.

248. **"On a Composite Model for Hadronic Constitutents,"** K. Matumoto, Prog. Theor. Phys. **52,** 1973–1975 (1974). (A)

proposed a model in which quarks are composed of a quartet of spin-$\frac{1}{2}$ particles carrying flavor, and a triplet of spin-0 particles carrying color, bound by an Abelian interaction.

249. **"Composite Models of Leptons,"** O. W. Greenberg and C. A. Nelson, Phys. Rev. D **10,** 2567–2573 (1974). (A)

gave criteria for composite models of leptons, and presented models which satisfy a number of the criteria.

*250. **"New Narrow Resonances and Separate Localization of Ordinary and Color SU(3),"** O. W. Greenberg, Phys. Rev. Lett. **35,** 1120–1123 (1975). (A)

suggested a model in which a quark is a composite of a spin-$\frac{1}{2}$, Fermi flavor-carrying object and a spin-0, Bose color-carrying object. A meson is then a four-quark system. An attempt was made to identify the J/ψ and associated resonances with excitations of this four-body system. References 248 and 250 seem to be the first to consider space excitations of subquark states and to consider subquarks as physical objects and not just as mathematical constructs.

*251. **"Are Quarks Composite?,"** J. C. Pati, A. Salam, and J. Strathdee, Phys. Lett. **59B,** 265–268 (1975). (A)

suggested models in which quarks and leptons are both composites of a valency-quartet Q which carries flavor and a color-quartet C which carries color, as well as, in some cases, a singlet spinor S. The preferred models have Q a fermion and C a boson, or Q, C, and S all fermions. They assigned two U(1) charges to the "preons" in such a way that the quarks and leptons are neutral.

Further interest in composite models was stimulated by the "rishon" model of Harari and a similar model of Shupe in which color is associated with the ordering of rishons in the three-rishon states which correspond to quarks: for example, if the rishons are T and V, then the states TTV, TVT, and VTT correspond to the three colors of the u quark,

*252. **"Schematic Model of Quarks and Leptons,"** H. Harari, Phys. Lett. B **86**, 83–86 (1979). (A)

*253. **"A Composite Model of Leptons and Quarks,"** M. A. Shupe, Phys. Lett. B **86**, 87–92 (1979). (A)

The rishon model does not appear to be compatible with the usual framework of quantum mechanics, because with the usual Bose or Fermi commutation relations for the rishon fields the three orderings which are supposed to correspond to color give linearly dependent states. J. C. Pati has pointed out (private communication) that there is no binding mechanism in the context of the rishon model which will produce the desired spectrum. Elaborations of the Harari–Shupe model have been proposed by several authors. Unfortunately, many of these elaborations lose the most striking feature of the model: the generation of color at the quark level from constituents which do not carry the color degree of freedom. These models include

254. **"A Dynamical Theory for the Rishon Model,"** H. Harari and N. Seiberg, Phys. Lett. **98B**, 269–273 (1981). (A)

Dynamical symmetry breaking in composite models, with particular application to the rishon model, is discussed by

255. **"Dynamical Symmetry Breaking and Composite Quarks and Leptons,"** H. Harari and N. Seiberg, Weizmann Institute preprint WIS-80/48-Nov.-Ph. (unpublished). (A)

A model closely related to the SO(10) grand unified theory in which, like the Harari–Shupe model, the non-Abelian symmetries such as color are not present at the constituent level was proposed by several authors, including

*256. **"Constituent Models for Basic Fields of Grand Unified Theories,"** F. Mansouri, Phys. Lett. **100B**, 25–28 (1981). (A)

Models in which quarks and leptons are composed entirely of the same constituents allow proton decay by constituent rearrangement. The proton lifetime comes out much too small in most such models; however, constraints on such models which ensure that the proton is sufficiently stable, while still having the observed quark and lepton spectrum, are given in

257. **"Search for Composite Models,"** R. Casalbuoni and R. Gatto, Univ. of Genève preprint UGVA-DPT 1981/06-279 (1981), pp. 1–89 (unpublished). (A)

't Hooft analyzed the constraints on composite models due to the absence of Adler–Bell–Jackiw anomalies, and to the Symanzik–Appelquist–Carrazone decoupling theorem, in

*258. **"Naturalness, Chiral Symmetry, and Spontaneous Chiral Symmetry Breaking,"** G. 't Hooft, in *Recent Developments in Gauge Theories*, edited by G. 't Hooft, C. Itzykson, A. Jaffe, H. Lehmann, P. K. Mitter, I. M. Singer, and S. Stora (Plenum, New York, 1980), pp. 135–157. (A)

't Hooft's analysis stimulated several other articles, including

259. **"Light Composite Fermions,"** S. Dimopoulos, S. Raby, and L. Susskind, Nucl. Phys. B **173**, 208 (1980). (A)

260. **"The Axial Anomaly and the Bound State Spectrum in Confining Theories,"** Y. Frishman, A. Schwimmer, T. Banks, and S. Yankielowicz, Nucl. Phys. B **177**, 157 (1981). (A)

Models in which the color, flavor, and generational degrees of freedom are carried by different constituents of quarks and leptons have been studied in several articles, including

261. **"Magnetism as the Origin of Preon Binding,"** J. C. Pati, Phys. Lett. **98B**, 40–44 (1981). (A)

262. **"A Preon Model with Hidden Electric and Magnetic Type Charges,"** J. C. Pati, Trieste International Centre for Theoretical Physics preprint IC/80/180 (1980), pp. 1–20 (unpublished). (A)

263. **"Subquark Model of Leptons and Quarks,"** H. Terazawa, Phys. Rev. D **22**, 184–199 (1980). (A)

The idea that the constituents of quarks, leptons, and other particles at that level may be bound by a new permanently confining force is appealing, and has been considered in Ref. 258 and related articles, as well as in

264. **"Subcomponent Models of Quarks and Leptons,"** R. Casalbuoni and R. Gatto, Phys. Lett. **93B**, 47–52 (1980). (A)

265. **"Algebra of Subquark Charges,"** H. Terazawa, Prog. Theor. Phys. **64**, 1973 (1980). (A)

266. **"Primitive Particle Model,"** Y. Ne'eman, Phys. Lett. B **82**, 69–70 (1979). (A)

267. **"A Quantum Structuredynamic Model of Quarks, Leptons, Weak Vector Bosons and Higgs Mesons,"** O. W. Greenberg and J. Sucher, Phys. Lett. **99B**, 339–343 (1981). (A)

The fact that the GIM mechanism (see Ref. 52) has a natural place in composite models was pointed out in

268. **"A Natural Composite Model for Quarks and Leptons,"** V. Visnjic-Triantafillou, FERMILAB-Pub-80/34-THY (1980), pp. 1–7 (unpublished); and in Refs. 265 and 267. (A)

Strong constraints on composite models follow from the great accuracy with which the measured magnetic moments of the electron and muon agree with predictions of quantum electrodynamics (QED) assuming point electrons and muons, as well as from the experimental upper limit on the decay muon to electron plus a photon. These issues are discussed in

269. **"Constraints on Composite Models of Quarks and Leptons,"** H. J. Lipkin, Phys. Lett. **103B**, 440–444 (1981). (A)

270. **"Magnetic Moments of Composite Fermions,"** G. L. Shaw, D. Silverman, and R. Slansky, Phys. Lett. **94B**, 57–60 (1980). (A)

271. **"Anomalous Magnetic Moment and Limits on Fermion Substructure,"** S. J. Brodsky and S. D. Drell, Phys. Rev. D **22**, 2236–2243 (1980). (A)

Numerical solution of a model in which a light, small, composite fermion is made from two heavy, relativistic, constituents bound by strong, short-range forces, while still obtaining the correct Dirac moment for the bound state, is given in

272. **"Dynamical Realization of Composite Leptons and Quarks,"** M. Bander, T. W. Chiu, G. L. Shaw, and D. Silverman, Fhys. Rev. Lett. **47**, 549–552; **Erratum,** *ibid.* **47**, 1419 (1981). (A)

Experimental signatures of composite structure of quarks at the lowest energy are of great interest. The suggestion that events occurring in pp, pp, ep, or e^+e^- collisions which have very large multiplicities of hadrons and leptons, with large transverse momenta and acoplanarity, might be an important signature occurs in

273. **"Glints: A Signature of Quark and Lepton Substructure,"** A. DeRujula, Phys. Lett. **96B**, 279–284 (1980). (A)

Grand unified theories such as SU(5) and SO(10) have a mass scale which is close to the Planck mass: nonetheless, they do not include the gravitational interaction. Extended supergravity theories attempt to provide a unified description of all interactions, including gravity; however, the maximal extended supergravity theory has too small an internal symmetry group [SO(8)] to contain the minimal SU(5) grand unified theory for the observed fermions. The suggestion that extended supergravity should be applied to preons, and that the observed fermions should be composite states of the preons was made and explored in several papers, including Ref. 247 above, and

274. **"A Grand Unified Theory Obtained from Broken Supergravity,"** J. Ellis, M. K. Gaillard, and B. Zumino, Phys. Lett. **94B,** 343–348 (1980). (A)

Reviews of issues and problems connected with composite models of quarks and leptons are given in Ref. 263 and in

*275. **"Compositeness of Quarks and Leptons,"** M. E. Peskin, Cornell preprint CLNS 81/516 (1981) (to appear in Proc. 1981 Int. Symposium of Lepton and Photon Interactions at High Energies. (A)

IX. QUARK SEARCHES AND QUARK CHEMISTRY

The question, Can isolated quarks or antiquarks [or diquarks, or objects carrying nonsinglet $SU(3)_{color}$ representations, generally] exist?, is fundamental for elementary particle physics. Even though the present theoretical consensus is that isolated quarks cannot exist, the experimental search for quarks is very important. All searches look for objects carrying fractional charge, in particular $+2/3$ and $\pm 1/3$, in units of the proton charge e. Although in the present standard model, objects which have fractional charge also carry color and should be confined, there are alternative possibilities: (1) quarks and other objects which carry color have integral charge, as discussed in an extensive series of articles by Pati and Salam and their collaborators. In addition to Refs. 223 and 224 above, a review of their program appears in

276. **"The Unconfined Unstable Quark, I and II,"** J. C. Pati and A. Salam, Comments Nucl. Part. Phys. **6,** 183–192 (1976); and **7,** 1–15 (1977). (I)
277. **"The Unification Puzzle,"** J. C. Pati, in *Proc. Seoul Symposium on Elementary Particle Physics in Memory of Benjamin W. Lee, 1978,* edited by J. Kim, P. Y. Pac, and H. S. Song (Seoul National University, Seoul, 1979), pp. 481–598. (A)

This possibility has also been considered by

278. **"Can Quarks Have Integer Charge?,"** B. Iijima and R. L. Jaffe, Phys. Rev. D **24,** 177–184 (1981). (A);

(2) confinement is not exact, and fractionally charged quarks can exist in isolation; (3) if fractionally charged particles are found, one or more of the following are true: they are not quarks, do not carry color, and do not interact strongly.

Quark searches fall into three categories: Searches with accelerators, searches in cosmic rays, and searches in stable matter. The literature up to 1977 is well reviewed in

279. **"A Review of Quark Search Experiments,"** L. W. Jones, Rev. Mod. Phys. **49,** 717–752 (1977). (A)

More recent work is reviewed in

280. **"Current Status of Quark Searches,"** L. Lyons, Oxford University preprint 38/80 (1980) (unpublished). (A)

The most exciting present development is the continuing collection of positive evidence for fractionally charged objects by the Stanford group. Their latest results are reported in

281. **"Observation of Fractional Charge of (1/3)*e* on Matter," G. S. LaRue, J. D. Phillips, and W. M. Fairbank, Phys. Rev. Lett. **46,** 967–970 (1981). (A)

The limits on quark abundance which can be inferred by the failure to detect quarks in a given type of stable matter depend crucially on quark chemistry; thus all stable matter searches must contain some discussion of quark chemistry. A specific discussion of quark chemistry is given in

282. **"The Chemistry of Free Quarks,"** K. S. Lackner and G. Zweig, California Institute of Technology preprint, CALT 68-781 (1980), pp. 1–19 (unpublished). (A)

X. QUARKS IN NUCLEAR PHYSICS

The quark model together with quantum chromodynamics should provide a fundamental basis for nuclear physics, just as the nuclear atom together with quantum electrodynamics provides a fundamental basis for atomic physics. The derivations of the nucleon–nucleon force and of other basic inputs for nuclear physics from quantum chromodynamics have not yet been achieved, despite the fact that the quark model was proposed in 1964; indeed, attempts to derive the nucleon–nucleon force started relatively recently. The most naive attempt uses the confining potential model, which, as mentioned in Sec. III, is successful in describing the properties of single hadrons. This attempt fails, because it leads to color-analog van der Waals forces which fall off as an inverse power of the distance between the hadrons, rather than exponentially as observed experimentally, for large separation between the hadrons, as discussed in Refs. 141–144 in Sec. V. It is possible that inclusion of effects, such as string breaking and pair production, which lie outside the potential model might allow the derivation of hadron–hadron interactions. These possibilities are mentioned in Ref. 144 and in

283. **"Permanent Quark Confinement Without van der Waals Forces,"** O. W. Greenberg and J. Hietarinta, Phys. Lett. **86B,** 309–312 (1979). (A)
284. **"Link-Operator Formulation of Quark Confinement Without van der Waals Forces,"** O. W. Greenberg and J. Hietarinta, Phys. Rev. D **22,** 993–998 (1980). (A)

A modification of the bag model which is popular for applications to nuclear physics uses a bag of smaller size than the MIT bag surrounded by a pion cloud outside the bag, the "little" bag or Brown–Rho bag model, reviewed in

285. **"The Little Bag and Nucleon–Nucleon Forces,"** G. E. Brown, in Proc. Workshop on Baryonium and Other Unusual Hadron States, Orsay, June, 1979 (unpublished). (A)

The one-boson-exchange model of hadron–hadron interactions is discussed on the basis of QCD in the context of the MIT bag model in

286. **"Bag-Model Quantum Chromodynamics for Hyperons at Low Energy,"** H. J. Weber and J. N. Maslow, Z. Phys. A **297,** 271–274 (1980). (A)

A review of the bag model with emphasis on the phenomenology of excited baryon states is given in

287. **"Phenomenological Aspects of the MIT Bag,"** A. J. G. Hey, Southampton University preprint 76/7-13 (1976), pp. 1–63; in Spanish Seminar (1977) (unpublished). (A)

XI. ASTROPHYSICAL AND COSMOLOGICAL IMPLICATIONS OF QUARKS

If the matter in the central core of a star has a density greater than the density of a neutron, which is about 8×10^{14} g/cm^{-3}, the neutrons in the core overlap, and it is possible that the clustering of three quarks per neutron disappears, and that the matter consists of a gas of quarks, rather than a gas of neutrons. The nature of quark matter and a possible transition between a degenerate neutron gas and a degenerate quark gas are reviewed in Ref. 27, p. 380.

Introductions to astrophysics and cosmology are given in

288. **The First Three Minutes—A Modern View of the Origin of the Universe,** S. Weinberg (Basic, New York, 1977). (E)
289. **Gravitation and Cosmology: Principles and Applications of the General Theory of Relativity,** S. Weinberg (Wiley, New York, 1972). (A)

It is a striking fact that the matter on the Earth, in the solar system, in our galaxy, and in our cluster of galaxies is composed of baryons, with a vanishingly small admixture of antibaryons. It is plausible to assume that there is an excess of baryons over antibaryons throughout the entire universe, as discussed in

290. **"Observational Tests of Antimatter Cosmologies,"** G. Steigman, Ann. Rev. Astron. Astrophys. **14,** 339–372 (1976). (A)

It was first pointed out in Ref. 230 that three conditions are necessary to generate a baryon excess from an initially symmetric big bang cosmology: (1) baryon number nonconserving interactions; (2) violation of *C* and *CP* (i.e., charge conjugation and charge conjugation–parity) symmetry; and (3) lack of thermal equilibrium.

Interest in generating the baryon excess was revived eleven years later by the article

*291. **"Unified Gauge Theories and the Baryon Number of the Universe,"** M. Yoshimura, Phys. Rev. Lett. **41,** 281–284 (1978). (A)

This question is also treated in

*292. **"The Cosmic Asymmetry Between Matter and Antimatter,"** F. Wilczek, Sci. Am. **243** (6), 82–102 (1980). (E)

293. **"Baryon Number of the Universe,"** S. Dimopoulos and L. Susskind, Phys. Rev. D **18,** 4500–4509 (1978). (A)

294. **"Matter–Antimatter Accounting, Thermodynamics, and Black-Hole Radiation,"** D. Toussaint, S. B. Treiman, F. Wilczek and A. Zee, Phys. Rev. D **19,** 1036–1045 (1979). (A)

295. **"Cosmological Production of Baryons,"** S. Weinberg, Phys. Rev. Lett. **42,** 850–853 (1979). (A)

A short review survey of cosmological baryon number generation appears in

296. **"Cosmological Baryon Number Generation in Grand Unified Models,"** J. A. Harvey, E. W. Kolb, D. B. Reiss, and S. Wolfram, Caltech preprint, CALT, 68-815 (unpublished). (A)

Some discussion of baryon asymmetry appears in most reviews on grand unified models, for example:

297. **"Grand Unification and Cosmology,"** J. Ellis, M. K. Gaillard, and D. V. Nanopoulos, in *Unification of the Fundamental Particle Interactions 1980,* edited by S. Ferrara, J. Ellis, and P. Van Nieuwenhuizen (Plenum, New York, 1980), pp. 461–493. (A)

ACKNOWLEDGMENTS

I am very happy to thank my colleagues Bill Caswell, Ashok Das, Tony Kennedy, Mike Ogilvie, Jogesh Pati, Joe Redish, Joe Sucher, and Ching-Hung Woo at the University of Maryland for reading drafts of this Resource Letter, and for giving me helpful suggestions about it. I am grateful to Delores Kight and Rosemary Robinson for editorial assistance as well as for typing the manuscript. This work was supported in part by a grant from the National Science Foundation and by a Faculty Research Grant from the University of Maryland.

A SCHEMATIC MODEL OF BARYONS AND MESONS *

M. GELL-MANN
California Institute of Technology, Pasadena, California

Received 4 January 1964

If we assume that the strong interactions of baryons and mesons are correctly described in terms of the broken "eightfold way" [1-3], we are tempted to look for some fundamental explanation of the situation. A highly promised approach is the purely dynamical "bootstrap" model for all the strongly interacting particles within which one may try to derive isotopic spin and strangeness conservation and broken eightfold symmetry from self-consistency alone [4]. Of course, with only strong interactions, the orientation of the asymmetry in the unitary space cannot be specified; one hopes that in some way the selection of specific components of the F-spin by electromagnetism and the weak interactions determines the choice of isotopic spin and hypercharge directions.

Even if we consider the scattering amplitudes of strongly interacting particles on the mass shell only and treat the matrix elements of the weak, electromagnetic, and gravitational interactions by means of dispersion theory, there are still meaningful and important questions regarding the algebraic properties of these interactions that have so far been discussed only by abstracting the properties from a formal field theory model based on fundamental entities [3] from which the baryons and mesons are built up.

If these entities were octets, we might expect the underlying symmetry group to be SU(8) instead of SU(3); it is therefore tempting to try to use unitary triplets as fundamental objects. A unitary triplet t consists of an isotopic singlet s of electric charge z (in units of e) and an isotopic doublet (u, d) with charges z+1 and z respectively. The anti-triplet $\bar{t}$ has, of course, the opposite signs of the charges. Complete symmetry among the members of the triplet gives the exact eightfold way, while a mass difference, for example, between the isotopic doublet and singlet gives the first-order violation.

For any value of z and of triplet spin, we can construct baryon octets from a basic neutral baryon singlet b by taking combinations $(b t \bar{t})$, $(b t t \bar{t} \bar{t})$, etc. **. From $(b t \bar{t})$, we get the representations **1** and **8**, while from $(b t t \bar{t} \bar{t})$ we get **1**, **8**, **10**, $\overline{\mathbf{10}}$, and **27**. In a similar way, meson singlets and octets can be made out of $(t \bar{t})$, $(t t \bar{t} \bar{t})$, etc. The quantum number $n_t - n_{\bar{t}}$ would be zero for all known baryons and mesons. The most interesting example of such a model is one in which the triplet has spin $\frac{1}{2}$ and $z = -1$, so that the four particles d^-, s^-, u^0 and b^0 exhibit a parallel with the leptons.

A simpler and more elegant scheme can be constructed if we allow non-integral values for the charges. We can dispense entirely with the basic baryon b if we assign to the triplet t the following properties: spin $\frac{1}{2}$, $z = -\frac{1}{3}$, and baryon number $\frac{1}{3}$. We then refer to the members $u^{\frac{2}{3}}$, $d^{-\frac{1}{3}}$, and $s^{-\frac{1}{3}}$ of the triplet as "quarks" [6] q and the members of the anti-triplet as anti-quarks $\bar{q}$. Baryons can now be constructed from quarks by using the combinations $(q q q)$, $(q q q q \bar{q})$, etc., while mesons are made out of $(q \bar{q})$, $(q q \bar{q} \bar{q})$, etc. It is assuming that the lowest baryon configuration $(q q q)$ gives just the representations **1**, **8**, and **10** that have been observed, while the lowest meson configuration $(q \bar{q})$ similarly gives just **1** and **8**.

A formal mathematical model based on field theory can be built up for the quarks exactly as for p, n, Λ in the old Sakata model, for example [3] with all strong interactions ascribed to a neutral vector meson field interacting symmetrically with the three particles. Within such a framework, the electromagnetic current (in units of e) is just

$$i\{\tfrac{2}{3}\bar{u}\gamma_\alpha u - \tfrac{1}{3}\bar{d}\gamma_\alpha d - \tfrac{1}{3}\bar{s}\gamma_\alpha s\}$$

or $\mathscr{F}_{3\alpha} + \mathscr{F}_{8\alpha}/\sqrt{3}$ in the notation of ref. [3]. For the weak current, we can take over from the Sakata model the form suggested by Gell-Mann and Lévy [7], namely $i\,\bar{p}\gamma_\alpha(1+\gamma_5)(n\cos\theta + \Lambda\sin\theta)$, which gives in the quark scheme the expression ***

$$i\,\bar{u}\gamma_\alpha(1+\gamma_5)(d\cos\theta + s\sin\theta)$$

* Work supported in part by the U.S. Atomic Energy Commission.

** This is similar to the treatment in ref. [1]. See also ref. [5].

*** The parallel with $i\,\bar{\nu}_e\gamma_\alpha(1+\gamma_5)e$ and $i\,\bar{\nu}_\mu\gamma_\alpha(1+\gamma_5)\mu$ is obvious. Likewise, in the model with d^-, s^-, u^0, and b^0 discussed above, we would take the weak current to be $i(\bar{b}^0\cos\theta + \bar{u}^0\sin\theta)\gamma_\alpha(1+\gamma_5)s^- + i(\bar{u}^0\cos\theta - \bar{b}^0\sin\theta)\gamma_\alpha(1+\gamma_5)d^-$. The part with $\Delta(n_t - n_{\bar{t}}) = 0$ is just $i\,\bar{u}^0\gamma_\alpha(1+\gamma_5)(d^-\cos\theta + s^-\sin\theta)$.

Volume 8, number 3 PHYSICS LETTERS 1 February 1964

or, in the notation of ref. 3),

$$[\mathcal{F}_{1\alpha} + \mathcal{F}^5_{1\alpha} + i(\mathcal{F}_{2\alpha} + \mathcal{F}^5_{2\alpha})] \cos\theta$$
$$+ [\mathcal{F}_{4\alpha} + \mathcal{F}^5_{4\alpha} + i(\mathcal{F}_{5\alpha} + \mathcal{F}^5_{5\alpha})] \sin\theta \, .$$

We thus obtain all the features of Cabibbo's picture [8] of the weak current, namely the rules $|\Delta I| = 1$, $\Delta Y = 0$ and $|\Delta I| = \frac{1}{2}$, $\Delta Y/\Delta Q = +1$, the conserved $\Delta Y = 0$ current with coefficient $\cos\theta$, the vector current in general as a component of the current of the F-spin, and the axial vector current transforming under SU(3) as the same component of another octet. Furthermore, we have [3] the equal-time commutation rules for the fourth components of the currents:

$$[\mathcal{F}_{j4}(x) \pm \mathcal{F}^5_{j4}(x), \mathcal{F}_{k4}(x') \pm \mathcal{F}^5_{k4}(x')] =$$
$$- 2 f_{jkl} \, [\mathcal{F}_{l4}(x) \pm \mathcal{F}^5_{l4}(x)] \, \delta(x-x') \, ,$$
$$[\mathcal{F}_{j4}(x) \pm \mathcal{F}^5_{j4}(x), \mathcal{F}_{k4}(x') \mp \mathcal{F}^5_{k4}(x')] = 0 \, ,$$

$i = 1, \ldots 8$, yielding the group SU(3) × SU(3). We can also look at the behaviour of the energy density $\theta_{44}(x)$ (in the gravitational interaction) under equal-time commutation with the operators $\mathcal{F}_{j4}(x') \pm \mathcal{F}_{j4}{}^5(x')$. That part which is non-invariant under the group will transform like particular representations of SU(3) × SU(3), for example like $(3, \bar{3})$ and $(\bar{3}, 3)$ if it comes just from the masses of the quarks.

All these relations can now be abstracted from the field theory model and used in a dispersion theory treatment. The scattering amplitudes for strongly interacting particles on the mass shell are assumed known; there is then a system of linear dispersion relations for the matrix elements of the weak currents (and also the electromagnetic and gravitational interactions) to lowest order in these interactions. These dispersion relations, unsubtracted and supplemented by the non-linear commutation rules abstracted from the field theory, may be powerful enough to determine all the matrix elements of the weak currents, including the effective strengths of the axial vector current matrix elements compared with those of the vector current.

It is fun to speculate about the way quarks would behave if they were physical particles of finite mass (instead of purely mathematical entities as they would be in the limit of infinite mass). Since charge and baryon number are exactly conserved, one of the quarks (presumably $u^{\frac{2}{3}}$ or $d^{-\frac{1}{3}}$) would be absolutely stable *, while the other member of the doublet would go into the first member very slowly by β-decay or K-capture. The isotopic singlet quark would presumably decay into the doublet by weak interactions, much as Λ goes into N. Ordinary matter near the earth's surface would be contaminated by stable quarks as a result of high energy cosmic ray events throughout the earth's history, but the contamination is estimated to be so small that it would never have been detected. A search for stable quarks of charge $-\frac{1}{3}$ or $+\frac{2}{3}$ and/or stable di-quarks of charge $-\frac{2}{3}$ or $+\frac{1}{3}$ or $+\frac{4}{3}$ at the highest energy accelerators would help to reassure us of the non-existence of real quarks.

These ideas were developed during a visit to Columbia University in March 1963; the author would like to thank Professor Robert Serber for stimulating them.

References

1) M.Gell-Mann, California Institute of Technology Synchrotron Laboratory Report CTSL-20 (1961).
2) Y.Ne'eman, Nuclear Phys. 26 (1961) 222.
3) M.Gell-Mann, Phys.Rev. 125 (1962) 1067.
4) E.g.: R.H.Capps, Phys.Rev. Letters 10 (1963) 312; R.E.Cutkosky, J.Kalckar and P.Tarjanne, Physics Letters 1 (1962) 93; E.Abers, F.Zachariasen and A.C .Zemach, Phys. Rev. 132 (1963) 1831; S.Glashow, Phys.Rev. 130 (1963) 2132; R.E.Cutkosky and P.Tarjanne, Phys.Rev. 132 (1963) 1354.
5) P.Tarjanne and V.L.Teplitz, Phys.Rev. Letters 11 (1963) 447.
6) James Joyce, Finnegan's Wake (Viking Press, New York, 1939) p.383.
7) M.Gell-Mann and M.Lévy, Nuovo Cimento 16 (1960) 705.
8) N.Cabibbo, Phys.Rev. Letters 10 (1963) 531.

* There is the alternative possibility that the quarks are unstable under decay into baryon plus anti-di-quark or anti-baryon plus quadri-quark. In any case, some particle of fractional charge would have to be absolutely stable.

* * * * *

VOLUME 13, NUMBER 20 PHYSICAL REVIEW LETTERS 16 NOVEMBER 1964

SPIN AND UNITARY-SPIN INDEPENDENCE IN A PARAQUARK MODEL OF BARYONS AND MESONS

O. W. Greenberg*
Institute for Advanced Study, Princeton, New Jersey
(Received 27 October 1964)

Wigner's supermultiplet theory,[1] transplanted independently by Gürsey, Pais, and Radicati,[2] and by Sakita,[2] from nuclear-structure physics to particle-structure physics, has aroused a good deal of interest recently. In the nuclear supermultiplet theory, the approximate independence of both spin and isospin of those forces relevant to the energies of certain low-lying bound states (nuclei) makes it useful to classify the states according to irreducible representations of SU(4). Parallel to this, in the particle supermultiplet theory, the possible independence of both spin and unitary spin of those forces relevant to the masses of certain low-lying bound states (particles) makes it interesting to classify the states according to irreducible representations of SU(6). Three results associated with this SU(6) classification indicate its usefulness: (1) The best known baryons (in particular, the spin-$\frac{1}{2}^+$ baryon octet and the spin-$\frac{3}{2}^+$ baryon decuplet) are grouped into a supermultiplet containing 56 particles.

VOLUME 13, NUMBER 20 PHYSICAL REVIEW LETTERS 16 NOVEMBER 1964

The pseudoscalar 0^- octet, vector 1^- octet, and vector 1^- singlet of mesons are grouped into a supermultiplet containing 35 particles. (2) Rather accurate mass formulas have been written down (partly on heuristic grounds) for the supermultiplets.[2,3] (3) In the approximation where the spin and unitary-spin independence of the forces relevant to the $\underline{56}$ is broken only by electromagnetic coupling, the magnetic moments of all the baryons in the $\underline{56}$ have been calculated up to a single common factor.[4] This calculation predicts the ratio $\beta = \mu_n/\mu_p = -\frac{2}{3}$, which agrees with the experimental value to within 3%.

Analogy with the Wigner supermultiplet theory leads us to adopt an atomic model in which all of the baryons and mesons are composite objects made up of basic particles. We assume that the basic forces are independent of unitary spin, but are in general spin dependent, and that the approximate symmetry group of the theory is $\mathcal{P}'\otimes SU(3)$, where $\mathcal{P}'$ is the covering group of the Poincaré group. However, for low-lying bound states, in particular when all orbital angular momenta are zero, all orbit-orbit and spin-orbit forces vanish so that there will be an additional degeneracy which will allow SU(6) invariance for these particular states, provided the spin-spin forces are not too large.[5] The desired supermultiplets, the $\underline{56}$ for the baryons and the $\underline{35}$ for the mesons, contain, respectively, the symmetric direct product of three SU(6) multiplets, and the antisymmetric direct product of an SU(6) and an SU(6)* multiplet. These assignments suggest that the basic particles are quarks.[6-9] For the mesons the $\underline{35}$ can be achieved by a state containing a fermion and an antifermion in s states; however, for the baryons the assignment of three fermions to the $\underline{56}$ is not possible if all three fermions are in s states.[10] We will return to this problem later, but for the moment assume that we can arrange to construct both the $\underline{35}$ and the $\underline{56}$ with all particles in s states.

The notion that an atomic model underlies the SU(6) invariance gives some hints for the derivation of mass formulas. In analogy with atomic and nuclear physics we suggest that the terms violating the exact SU(6) symmetry, for those states where the symmetry is relevant, arise either from one-body or two-body forces. A particularly simple assumption is that all the SU(6)-violating terms come from the $J=I=Y=0$ member of the (35-dimensional) adjoint representation of SU(6). With this assumption, the one-body force gives for the mass operator

$$\mathfrak{M}^{(1)} = \sum_i M_0 V_0(r_i) + \sum_i [T_3^{\,3}(i) + T_6^{\,6}(i)] V_1(r_i),$$

where $T_i^{\,j}$ is a tensor operator which transforms in the same way as the adjoint representation. The two-body contribution gives

$$\mathfrak{M}^{(2)} = \tfrac{1}{2} \sum_{i\neq j} \sum_{\lambda} [(T_3^{\,3} + T_6^{\,6})_i \times (T_3^{\,3} + T_6^{\,6})_j]^{(\lambda)} V_2^{(\lambda)}(|r_i - r_j|),$$

where λ distinguishes the different contributions which can occur from the two-body operator in the brackets. These contributions are $M_{(1)}^{(1)}$, $M_{(35)}^{(8)}$, $M_{(405)}^{(1)}$, $M_{(405)}^{(8)}$, and $M_{(405)}^{(27)}$, using the notation of Bég and Singh.[3] Note that with this assumption, only the representations $\underline{1}$, $\underline{35}$, and $\underline{405}$ can contribute to $M^{(2)}$, because these are the only symmetric real representations contained in $\underline{35}\otimes\underline{35}$. For the $\underline{56}$, the contributions from $M_{(1)}^{(1)}$ and $M_{(35)}^{(8)}$ can be absorbed in $\mathfrak{M}^{(1)}$. Contributions from $M_{(405)}^{(1)}$ and $M_{(405)}^{(8)}$, together with the one-body operator, already suffice to give the mass formula which has been suggested for the $\underline{56}$:

$$M = M_0 + \alpha Y + \beta[I(I+1) - \tfrac{1}{4}Y^2] + \gamma J(J+1).$$

The magnetic-moment calculation,[4] as already pointed out by the authors, can be obtained by assuming that the entire magnetic moments of the baryons in the $\underline{56}$ are produced by the intrinsic magnetic moments of the quarks,[11] and that the quark moments are proportional to their Dirac moments, $Q\hbar/2Mc$, Q = quark charge, M = quark mass. If the free quark moments equal their Dirac moments, then one can resolve the incompatibility of assumptions (I)-(IV) discussed by Bég, Lee, and Pais[4] by saying that quarks, rather than nucleons, have minimal electromagnetic coupling, and that the radiative corrections (in empty space) to the quark moments are small (as is the case for the electron and muon), or proportional to their Dirac moments. However, if the quarks bound in nucleons have Dirac moments, then $M = m_p/2.79 = 336$ MeV,[12] so that one must look for a mechanism (perhaps acting only in strongly bound states) which enhances quark magnetic moments in proportion to their Dirac moments.

Now we return to the question of placing three spin-$\frac{1}{2}$ quarks in s states in the baryon $\underline{56}$.

VOLUME 13, NUMBER 20 PHYSICAL REVIEW LETTERS 16 NOVEMBER 1964

This can be done if the quarks are parafermions of order $p=3$. This suggestion is the main new idea of this article.[13] For parafermions of order 3, one has the Green Ansatz for the creation and annihilation operators,

$$a_\lambda^\dagger = \sum_{\alpha=1}^{3} a_\lambda^{(\alpha)\dagger}, \quad a_\lambda = \sum_{\alpha=1}^{3} a_\lambda^{(\alpha)},$$

where the $a^{(\alpha)}$ and $a^{(\beta)\dagger}$ satisfy the anticommutation rules for the same α,

$$[a_\lambda^{(\alpha)}, a_\mu^{(\alpha)\dagger}]_+ = \delta_{\lambda,\mu}, \quad [a_\lambda^{(\alpha)\dagger}, a_\mu^{(\alpha)\dagger}]_+ = 0,$$

and commute for different α and β,

$$[a_\lambda^{(\alpha)}, a_\mu^{(\beta)\dagger}]_- = 0, \quad \alpha \neq \beta;$$

$$[a_\lambda^{(\alpha)\dagger}, a_\mu^{(\beta)\dagger}]_- = 0, \quad \alpha \neq \beta.$$

Here λ stands for the single-particle quantum numbers; for example, momentum, spin, I_z, and Y. Let

$$f_{\lambda\mu\nu}^\dagger \equiv [[a_\lambda^\dagger, a_\mu^\dagger]_+, a_\nu^\dagger]_+$$

$$= 4 \sum_{\substack{\alpha,\beta,\gamma=1 \\ \alpha\neq\beta\neq\gamma\neq\alpha}}^{3} a_\lambda^{(\alpha)\dagger} a_\mu^{(\beta)\dagger} a_\nu^{(\gamma)\dagger}$$

Then the state $f_{\lambda\mu\nu}^\dagger \Phi_0$ is symmetric under all permutations of λ, μ, and ν.[14] This composite state is a fermion,[15] since $[f_{\lambda\mu\nu}^\dagger, a_\sigma^\dagger]_+ = 0$, which implies $[f_{\lambda\mu\nu}^\dagger, f_{\sigma\tau\eta}^\dagger]_+ = 0$.

The comparison of superselection sectors for paraquarks due to their para nature with the charge and baryon-number superselection sectors is given in Table I.

The suggestion that quarks are parafermions (and, in a field theory, the quanta of para-Fermi fields) is allowed by the selection rules which follow from locality.[13] Relevant[16] local interactions (in the sense of spacelike commutativity of the interaction Hamiltonian density) can be constructed with these fields; for example, the Yukawa interaction

$$H_I = g[[\bar{\Psi}, \Psi]_+ - \langle[\bar{\Psi}, \Psi]_+\rangle_0, \varphi]_+,$$

or the Fermi interaction

$$H_I = G\{[[\bar{\Psi}, \Psi]_-, [\bar{\Psi}, \Psi]_-]_+ - \langle\text{same}\rangle_0\},$$

where Ψ and φ are, respectively, para-Fermi and para-Bose of order 3. The paraquark suggestion does not fall under the quantum-mechanical theorem[17] that particles with anomalous permutation properties cannot be produced from initial states ($\mathfrak{F}^\times$) with at most one such particle because, in this theorem, it was assumed that the only superselection rules were those generated by charge, baryon number, and lepton number, while in parafield theories there are additional superselection rules.[13,18] The question of the compatibility of parafield theory with intuitive notions concerning the behavior of quantum systems when separated into subsystems has been raised.[19] This question deserves serious attention; however, we do not consider it here.

Table I. Comparison of para, charge, and baryon number superselection sectors for $p=3$ para-Fermi quarks.

State[a]	Para superselection sector	Q	B
$a^\dagger\Phi_0$	para-Fermi	$\frac{2}{3}, -\frac{1}{3}$	$\frac{1}{3}$
$[a^\dagger, a^\dagger]_+\Phi_0$	para-Bose	$\frac{4}{3}, \frac{1}{3}, -\frac{2}{3}$	$\frac{2}{3}$
$[a^\dagger, b^\dagger]_+\Phi_0$	para-Bose	$1, 0, -1$	0
$[a^\dagger, a^\dagger]_-\Phi_0$	Bose	$\frac{4}{3}, \frac{1}{3}, -\frac{2}{3}$	$\frac{2}{3}$
$[a^\dagger, b^\dagger]_-\Phi_0$	Bose	$1, 0, -1$	0
$[[a^\dagger, a^\dagger]_+, a^\dagger]_+\Phi_0$	Fermi	$2, 1, 0, -1$	1
$[[a^\dagger, a^\dagger]_+, b^\dagger]_+\Phi_0$	Fermi	$5/3, \frac{2}{3}, -\frac{1}{3}, -\frac{4}{3}$	$\frac{1}{3}$

[a] $a^\dagger$ is the creation operator for a quark, $b^\dagger$ for an antiquark.

After this brief defense of the possibility that quarks are para-Fermi, we proceed to the classification of baryon states in the paraquark model. The baryon wave functions must be symmetric under permutations. We present in Table II states in which the wave functions[5] are sums of products of space wave functions and spin–unitary-spin wave functions. All the quarks are in the lowest radial state for a given l. The Young diagrams for both the space and spin–unitary-spin wave functions are the same; the numbers listed give the number of boxes in successive rows. Even if the SU(6) classification is useful only for the lowest $\underline{56}$ with configuration s^3, the (SU(3), J) classification given in column 5 should be useful for higher states. We have continued the table up to the p^3 configuration in order to obtain states with $J=\frac{9}{2}$, since there is some experimental evidence for such states.[20] However, we postpone assigning the known baryons to multiplets. The fact that only the $\underline{1}$, $\underline{8}$, and $\underline{10}$ SU(3) mul-

VOLUME 13, NUMBER 20 PHYSICAL REVIEW LETTERS 16 NOVEMBER 1964

Table II. Low-lying states in paraquark model of baryons.[a]

Orbitals configuration	L	Parity	Young diagram	(SU(3), J) decomposition	Total multiplicity	Total no. of I multiplets
s^3 (pure)	0	+	(3)	$(\underline{8}, J=1/2)$ $(\underline{10}, J=3/2)$	56	8
s^2p^1 (spurious)	1	−	(3)	$(\underline{8}, J=1/2, 3/2)$ $(\underline{10}, J=1/2, 3/2, 5/2)$	168	20
s^2p^1 (pure)	1	−	(2, 1)	$(\underline{1}, J=1/2, 3/2)$ $(\underline{8}, J=1/2, 3/2, 5/2)$ $(\underline{10}, J=1/2, 3/2)$ $(\underline{8}, J=1/2, 3/2)$	210	30
s^1p^2, s^2d^1 (mixed)	2	+	(3)	$(\underline{8}, J=3/2, 5/2)$ $(\underline{10}, J=1/2, 3/2, 5/2, 7/2)$	280	24
s^1p^2, s^2d^1 (mixed)	2	+	(2, 1)	$(\underline{1}, J=3/2, 5/2)$ $(\underline{8}, J=1/2, 3/2, 5/2, 7/2)$ $(\underline{10}, J=3/2, 5/2)$ $(\underline{8}, J=3/2, 5/2)$	350	34
s^1p^2, s^2d^1 (mixed)	1	+	(2, 1)	$(\underline{1}, J=1/2, 3/2)$ $(\underline{8}, J=1/2, 3/2, 5/2)$ $(\underline{10}, J=1/2, 3/2)$ $(\underline{8}, J=1/2, 3/2)$	210	30
s^1p^2 (pure)	1	+	(1, 1, 1)	$(\underline{8}, J=1/2, 3/2)$ $(\underline{1}, J=1/2, 3/2, 5/2)$	60	11
s^1p^2, s^2d^1 (mixed)	0	+	(3)	$(\underline{8}, J=1/2)$ $(\underline{10}, J=3/2)$	56	8
s^1p^2, s^2d^1 (mixed)	0	+	(2, 1)	$(\underline{1}, J=1/2)$ $(\underline{8}, J=3/2)$ $(\underline{10}, J=1/2)$ $(\underline{8}, J=1/2)$	70	13
$p^3, s^1p^1d^1, s^2f^1$ (mixed)	3	−	(3)	$(\underline{8}, J=5/2, 7/2)$ $(\underline{10}, J=3/2, 5/2, 7/2, 9/2)$	392	24
$p^3, s^1p^1d^1$ (mixed)	2	−	(2, 1)	$(\underline{1}, J=3/2, 5/2)$ $(\underline{8}, J=1/2, 3/2, 5/2, 7/2)$ $(\underline{10}, J=3/2, 5/2)$ $(\underline{8}, J=3/2, 5/2)$	350	34
$p^3, s^1p^1d^1, s^2f^1$ (mixed)	1	−	(3)	$(\underline{8}, J=1/2, 3/2)$ $(\underline{10}, J=1/2, 3/2, 5/2)$	168	20
$p^3, s^1p^1d^1$ (mixed)	1	−	(2, 1)	$(\underline{1}, J=1/2, 3/2)$ $(\underline{8}, J=1/2, 3/2, 5/2)$ $(\underline{10}, J=1/2, 3/2)$ $(\underline{8}, J=1/2, 3/2)$	210	30
$p^3, s^1p^1d^1$ (mixed)	0	−	(1, 1, 1)	$(\underline{8}, J=1/2)$ $(\underline{1}, J=3/2)$	20	5
				Totals:	2600	291

[a]See reference 23.

tiplets occur is common to any three-quark model. The table can be constructed using only elementary facts[21] known to nuclear physicists. A useful check[22] on the construction of the table is the observation that if the s- and p-orbital states are considered equivalent, then these four states transform as the fundamental representation of an "orbital" SU(4) group. When this last group is combined with SU(6), one reaches SU(24), whose three-particle symmetric representation has dimension 2600.[23]

If this model is correct, then it should be possible to produce real paraquarks in high-energy interactions. The superselection rules for production of paraquarks from normal matter are the same as for Fermi quarks (see Table I); for example, $b+m \to 3q$, but $b+m \not\to 2q$

VOLUME 13, NUMBER 20 PHYSICAL REVIEW LETTERS 16 NOVEMBER 1964

or q, where b, m, q stand for baryon, meson, quark. However, the threshold behavior of $b+m \to 3q$ for paraquarks would reflect the s-state wave functions and would differ from the threshold behavior for Fermi quarks.

In Coulomb scattering from normal matter, the lowest order cross section for para particles is the same as for normal particles of the same charge, so para particles should be detected as easily as normal particles.

We thank Professor J. Robert Oppenheimer for his warm hospitality at the Institute for Advanced Study. We have benefited from the atmosphere in the physics community here. We thank, in particular, A. J. Dragt, P. G. O. Freund, B. W. Lee, and S. MacDowell for helpful discussions.

---

*Alfred P. Sloan Foundation Fellow on leave from the University of Maryland, College Park, Maryland.

[1]E. P. Wigner, Phys. Rev. 51, 106 (1937); E. P. Wigner and E. Feenberg, Rept. Progr. Phys. 8, 274 (1941).

[2]F. Gürsey and L. A. Radicati, Phys. Rev. Letters 13, 173 (1964); A. Pais, Phys. Rev. Letters 13, 175 (1964); F. Gürsey, A. Pais, and L. A. Radicati, Phys. Rev. Letters 13, 299 (1964); B. Sakita, Phys. Rev. (to be published).

[3]T. K. Kuo and T. Yao, Phys. Rev. Letters 13, 415 (1964); M. A. B. Bég and V. Singh, Phys. Rev. Letters 13, 418 (1964).

[4]M. A. B. Bég, B. W. Lee, and A. Pais, Phys. Rev. Letters 13, 514 (1964); B. Sakita, to be published.

[5]We have no theoretical justification for the use of a nonrelativistic classification of forces and states.

[6]M. Gell-Mann, Phys. Letters 8, 214 (1964); G. Zweig, to be published.

[7]B. Sakita, reference 2.

[8]P. G. O. Freund and B. W. Lee (private communication) have constructed models of baryons using Fermi quarks.

[9]It is amusing that one reaches the concrete notion of atomic structure after a lengthy detour through the special unitary groups.

[10]This was already pointed out by Sakita,[2] who suggested for this reason the use of the antisymmetric 20-dimensional representation for the baryons. However, this suggestion has two defects: First, the 20 no longer accommodates the decuplet; and secondly, the SU(6)-invariant electromagnetic coupling to the 20 leads to the wrong value of β.[4]

[11]This will be the case if the quarks are all in s states.

[12]Quarks having this mass would already have been detected. See W. Blum et al., Phys. Rev. Letters 13, 353a (1964); V. Hagopian et al., Phys. Rev. Letters 13, 280 (1964); and references in these articles.

[13]All statements made here concerning para particles are straightforward consequences of results contained in O. W. Greenberg and A. M. L. Messiah, to be published. The "absence of para particles" in the title of this article refers to the presently known particles.

[14]Permutations of these indices are related to permutations of the particles in the state for one-dimensional representations of the permutation group, but not in general.[13]

[15]We can also construct the meson 35 using a particle-antiparticle pair of order 3 para-Fermi quarks in s states. Let

$$b_{\lambda\mu}^{\dagger} = [a_{\lambda}^{\dagger}, a_{\mu}^{\dagger}]_{-} = 2 \sum_{\alpha \neq \beta = 1}^{3} a_{\lambda}^{(\alpha)\dagger} a_{\mu}^{(\beta)\dagger}.$$

Then the state $b_{\lambda\mu}^{\dagger}\Phi_0$ is antisymmetric under permutation of λ and μ, and this composite state is a boson since $[b_{\lambda\mu}^{\dagger}, a_{\nu}^{\dagger}]_{-} = 0$, which implies $[b_{\lambda\mu}^{\dagger}, b_{\nu\sigma}^{\dagger}]_{-} = 0$. This construction works for all para orders, including $p = 1$, the Fermi case. So the mesons constructed with parafermions are the same as those constructed with fermions; for this reason we will not consider the mesons further in this article.

[16]By "relevant," we mean that these interactions lead to connected Feynman graphs with only the desired 56 (or 35) in the initial and final states.

[17]A. M. L. Messiah and O. W. Greenberg, Phys. Rev. 136, B248 (1964).

[18]We expect that when this theorem[17] is amended to include the parafield superselection rules, the absolute selection rule for production of paraquarks from $\mathfrak{F}^{\times}$ will allow production, from ordinary states, of both symmetric and antisymmetric paraquark states, provided these states occur in different superselection sectors.

[19]C. N. Yang, private communication.

[20]A. H. Rosenfeld et al., Rev. Mod. Phys. (to be published).

[21]A. de-Shalit and I. Talmi, Nuclear Shell Theory (Academic Press, Inc., New York, 1963).

[22]We thank P. G. O. Freund and B. W. Lee for pointing this out.

[23]Because Table II lists shell-model states, some spurious states are present due to incorrect treatment of the center-of-mass motion. Only the configuration s^2p^1, with Young diagram (3), is purely spurious; for the other configurations we have indicated the higher shell-model states which must be mixed in to produce proper internal states. (Harmonic oscillator wave functions were used.) We plan to do more detailed calculations with more specific assumptions about the forces to try to estimate where the various SU(3) multiplets should lie. Perhaps other orbitals, such as the second radial s state, should be considered. Empirical information will be helpful in guiding such calculations. See J. P. Elliott and T. H. R. Skyrme, Proc. Roy. Soc. (London) A232, 561 (1955), and E. Baranger and C. W. Lee, Nucl. Phys. 22, 157 (1961), for discussions of spurious states.

Three-Triplet Model with Double $SU(3)$ Symmetry*

M. Y. HAN
Department of Physics, Syracuse University, Syracuse, New York

AND

Y. NAMBU
The Enrico Fermi Institute for Nuclear Studies, and the Department of Physics, The University of Chicago, Chicago, Illinois

(Received 12 April 1965)

With a view to avoiding some of the kinematical and dynamical difficulties involved in the single-triplet quark model, a model for the low-lying baryons and mesons based on three triplets with integral charges is proposed, somewhat similar to the two-triplet model introduced earlier by one of us (Y. N.). It is shown that in a $U(3)$ scheme of triplets with integral charges, one is naturally led to three triplets located symmetrically about the origin of I_3-Y diagram under the constraint that the Nishijima–Gell-Mann relation remains intact. A double $SU(3)$ symmetry scheme is proposed in which the large mass splittings between different representations are ascribed to one of the $SU(3)$, while the other $SU(3)$ is the usual one for the mass splittings within a representation of the first $SU(3)$.

I. INTRODUCTION

ALTHOUGH the $SU(6)$ symmetry strongly indicates that the baryon is essentially a three-body system built from some basic triplet field or fields, the quark model[1] is not entirely satisfactory from a realistic point of view, because (a) the electric charges are not integral, (b) three quarks in s states do not form the symmetric $SU(6)$ representation assigned to the baryons, and (c) a simple dynamical mechanism is lacking for realizing only zero-triality states as the low-lying levels.

These difficulties may be avoided if we introduce more than one basic triplet. Recently one of us (Y. N.) has attempted a two-triplet model[2] where the members of the triplets t_1 and t_2 had the charge assignment (1,0,0) and (0, −1, −1), as had been proposed earlier by Bacry *et al.*[3] The baryon would be represented by the combination $t_1t_1t_2$, whereas the mesons would correspond to some combination $\sim at_1\bar{t}_1'+bt_2\bar{t}_2'$. The triplets are assumed to have masses large compared to the baryon mass, which would mean that baryons and mesons have very large binding energies. A dynamical mechanism for this is provided by a neutral field coupled strongly to the "charm number"[4] C, which is 1 for t_1 and −2 for t_2, and therefore $C=0$ for baryons and mesons. In analogy with electrostatic energy, we can argue that the potential energy due to the charm field would be lowest when the system is "neutral," namely, $C=0$. Thus all other unwanted configurations with $C\neq 0$, which include among others triplet, sextet, etc. representations, would have high masses, and hence would not be easily observed.

There have been proposed two different ways in which to introduce basic triplet or triplets with integral charges. One approach essentially involves a modification of the Nishijima–Gell-Mann relation by way of introducing an additional quantum number, the triality quantum number,[5] and this has led to considerations of higher symmetry schemes based on rank-three Lie groups.[6] On the other hand, Okubo *et al.*[7] have recently shown that the minimal group required for this purpose is actually the group $U(3)$.[8] It is shown that a triplet scheme may be defined in $U(3)$ such that the triplet always possesses integral values of charge and hypercharge and satisfies the Nishijima–Gell-Mann relation without a modification. The $U(3)$ triplet considered by Okubo *et al.* is of Sakata type; i.e., it consists of an isodoublet and an isosinglet. Actually, the $U(3)$ scheme is much more appealing than those of the rank-three Lie groups on two accounts: firstly, the Nishijima–Gell-Mann relation is satisfied universally by triplets as by octets and decuplets, and secondly as far as the hitherto realized representations are concerned, $U(3)$ is equivalent to $SU(3)$.[9]

In what follows, we show that the $U(3)$ scheme, when fully utilized as described below, naturally and uniquely

* Work supported in part by the U. S. Atomic Energy Commission under the Contract No. AT(30-1)-3399 and No. AT(11-1)-264.

[1] M. Gell-Mann, Phys. Letters **8**, 214 (1964); G. Zweig, CERN (to be published).

[2] Y. Nambu, *Proceedings of the Second Coral Gables Conference on Symmetry Principles at High Energy* (W. H. Freeman and Company, San Francisco, 1965).

[3] H. Bacry, J. Nuyts, and L. van Hove, Phys. Letters **9**, 279 (1964).

[4] This name was originally used in connection with the $SU(4)$ symmetry. B. J. Bjørken and S. L. Glashow, Phys. Letters **11**, 255 (1964); A. Salam, Dubna Conference Report, 1964 (unpublished).

[5] G. E. Baird and L. C. Biedenharn, *Proceedings of the First Coral Gables Conference on Symmetry Principles at High Energy* (W. H. Freeman and Company, San Francisco, 1964); C. R. Hagen and A. J. Macfarlane, Phys. Rev. **135**, B432 (1964) and J. Math. Phys. **5**, 1335 (1964).

[6] For example, see I. S. Gerstein and M. L. Whippmann, Phys. Rev. **137**, B1522 (1965). Earlier references are given in this paper.

[7] S. Okubo, C. Ryan, and R. E. Marshak, Nuovo Cimento **34**, 759 (1964).

[8] The use of $U(3)$ in this connection has also been remarked by I. S. Gerstein and K. T. Mahanthappa, Phys. Rev. Letters **12**, 570, 656(E) (1964).

[9] S. Okubo, Phys. Letters **4**, 14 (1963).

B 1007 THREE-TRIPLET MODEL WITH DOUBLE $SU(3)$ SYMMETRY

leads to a set of three basic triplets with integral charges, namely an I-triplet (isodoublet and isosinglet), a U-triplet (U-spin doublet and U-spin singlet) and a V-triplet (V-spin doublet and V-spin singlet).[10] These triplets arise from three different ways of defining charge Q, hypercharge Y, and a displaced isospin I_3 in the $U(3)$ group as opposed to the $SU(3)$, in such a way that the charge and hypercharge have integral values, while keeping the Nishijima–Gell-Mann relation intact, and they differ from each other in their quantum-number assignments as well as in their transformation properties under the Weyl reflections.[11] This is described in Sec. II. In Sec. III, a double $SU(3)$ symmetry scheme is proposed based on the three-triplet model in which the large mass splittings between different representations are ascribed to one of the $SU(3)$, and the other $SU(3)$ is, as usual, responsible for the mass splittings within a representation. The low-lying baryon and meson states may be taken as singlets with respect to one of the $SU(3)$. The extended symmetry group with respect to the $SU(6)$ symmetry is briefly discussed.

II. THREE TRIPLETS

We shall denote the infinitesimal generators of $U(3)$ by $A_\nu{}^\mu$ which satisfies the following commutation relations:

$$[A_\beta{}^\alpha, A_\nu{}^\mu] = \delta_\beta{}^\mu A_\nu{}^\alpha - \delta_\nu{}^\alpha A_\beta{}^\mu, \tag{1}$$

where all indices take on the values 1, 2, and 3. The corresponding infinitesimal generators $B_\nu{}^\mu$ of $SU(3)$ are then given by

$$B_\nu{}^\mu = A_\nu{}^\mu - \tfrac{1}{3}\delta_\nu{}^\mu A_\lambda{}^\lambda \tag{2}$$

which satisfy the following equations:

$$[B_\beta{}^\alpha, B_\nu{}^\mu] = \delta_\beta{}^\mu B_\nu{}^\alpha - \delta_\nu{}^\alpha B_\beta{}^\mu \tag{3}$$

and

$$B_\lambda{}^\lambda = 0. \tag{4}$$

Furthermore, the unitary restriction gives

$$(A_\nu{}^\mu)^\dagger = A_\mu{}^\nu, \quad (B_\nu{}^\mu)^\dagger = B_\mu{}^\nu. \tag{5}$$

Let us now briefly summarize the relevant results of Okubo *et al.* In the $SU(3)$ scheme, the charge Q, the hypercharge Y and the third component of isospin I_3 are identified as follows[12]:

$$Q = -B_1{}^1, \tag{6a}$$

$$Y = B_3{}^3 = -B_1{}^1 - B_2{}^2 \quad [\text{by the relation (4)}], \tag{6b}$$

$$I_3 = \tfrac{1}{2}(B_2{}^2 - B_1{}^1). \tag{6c}$$

In the $U(3)$ scheme, the corresponding quantities $\tilde{Q}$, $\tilde{Y}$, and $\tilde{I}_3$ are defined as follows:

$$\tilde{Q} = -A_1{}^1 = Q - \tfrac{1}{3}\tau, \tag{7a}$$

$$\tilde{Y} = -A_1{}^1 - A_2{}^2 = Y - \tfrac{2}{3}\tau, \tag{7b}$$

$$\tilde{I}_3 = \tfrac{1}{2}(A_2{}^2 - A_1{}^1) = I_3, \tag{7c}$$

where

$$\tau = A_1{}^1 + A_2{}^2 + A_3{}^3. \tag{8}$$

With these definitions, the Nishijima–Gell-Mann relation is seen to be equally satisfied by the $U(3)$ and $SU(3)$ theories, i.e.,

$$Q = I_3 + \tfrac{1}{2}Y \tag{9}$$

and

$$\tilde{Q} = \tilde{I}_3 + \tfrac{1}{2}\tilde{Y}, \tag{10}$$

respectively. Since the generators $A_1{}^1$, $A_2{}^2$, and $A_3{}^3$ possess integral eigenvalues in any representation,[13] the identifications of $\tilde{Q}$ and $\tilde{Y}$ to be the charge and the hypercharge, respectively, in $U(3)$ theory shall always lead to integral values for the charge and the hypercharge. In particular, in the three-dimensional representation, the $U(3)$ triplet has the eigenvalues

$$\tilde{Q} = \begin{bmatrix} 1 & 0 & 0 \\ 0 & 0 & 0 \\ 0 & 0 & 0 \end{bmatrix}, \quad \tilde{I}_3 = \begin{bmatrix} \frac{1}{2} & 0 & 0 \\ 0 & -\frac{1}{2} & 0 \\ 0 & 0 & 0 \end{bmatrix}, \quad \tilde{Y} = \begin{bmatrix} 1 & 0 & 0 \\ 0 & 1 & 0 \\ 0 & 0 & 0 \end{bmatrix}. \tag{11}$$

This triplet corresponds to the Sakata triplet which we call an I triplet for short.

We can now generalize the above constructions of the $U(3)$ triplet in the following way. Comparing (6b) and (7b), we see that a particular choice has been made for $\tilde{Y}$. Had we defined $\tilde{Y}$ to be $A_3{}^3$, it would still have integral eigenvalues but the relation (10) would have been violated. This is because $B_\lambda{}^\lambda = 0$ in $SU(3)$ but $A_\lambda{}^\lambda \neq 0$ in general in $U(3)$ and thus some care is needed in defining corresponding quantities in $U(3)$. Making use of (4), the definition in (6) can be written more generally as

$$Q = -B_1{}^1 = B_2{}^2 + B_3{}^3, \tag{12a}$$

$$Y = B_3{}^3 = -B_1{}^1 - B_2{}^2, \tag{12b}$$

$$I_3 = \tfrac{1}{2}(B_2{}^2 - B_1{}^1) = \tfrac{1}{2}(2B_2{}^2 + B_3{}^3) = -\tfrac{1}{2}(2B_1{}^1 + B_3{}^3). \tag{12c}$$

As in (7), replacing $B_\nu{}^\mu$'s in (12) by corresponding $A_\nu{}^\mu$'s, we list all possible candidates for the corresponding quantities in $U(3)$ which are now however not equivalent to each other [they are equivalent, of course, when reduced to $SU(3)$], i.e.,

$$\tilde{Q}: \quad -A_1{}^1, \quad A_2{}^2 + A_3{}^3, \tag{13a}$$

$$\tilde{Y}: \quad A_3{}^3, \quad -A_1{}^1 - A_2{}^2, \tag{13b}$$

$$\tilde{I}_3: \quad \tfrac{1}{2}(A_2{}^2 - A_1{}^1), \quad \tfrac{1}{2}(2A_2{}^2 + A_3{}^3), \quad -\tfrac{1}{2}(2A_1{}^1 + A_3{}^3). \tag{13c}$$

---

[10] C. A. Levinson, H. J. Lipkin, and S. Meshkov, Nuovo Cimento **23**, 236 (1961); Phys. Letters **1**, 44 (1962) and Phys. Rev. Letters **10**, 361 (1963).

[11] A. J. Macfarlane, E. C. G. Sudarshan, and C. Dullemond, Nuovo Cimento **30**, 845 (1963).

[12] We use the sign convention of S. P. Rosen, J. Math. Phys. **5**, 289 (1964).

[13] For a derivation of this result, see Eq. (7) of Ref. 7.

To start with, the alternative choices in (13) provide twelve inequivalent ways in which to choose a set of three quantities $\tilde{Q}$, $\tilde{Y}$ and $\tilde{I}_3$ for the $U(3)$ scheme. In every choice $\tilde{Q}$ and $\tilde{Y}$ will have integral eigenvalues, but as can be easily checked the Nishijima–Gell-Mann relation will not be valid for all of them. In fact, there are only three cases for which it is valid and we are thus naturally led to three inequivalent triplets in the $U(3)$ scheme; they are defined by the following three choices:

$$t_I:\quad \tilde{Q}=-A_1{}^1,\qquad \tilde{Y}=-A_1{}^1-A_2{}^2,\qquad \tilde{I}_3=\tfrac{1}{2}(A_2{}^2-A_1{}^1),\tag{14a}$$

$$t_U:\quad \tilde{Q}=A_2{}^2+A_3{}^3,\qquad \tilde{Y}=A_3{}^3,\qquad \tilde{I}_3=\tfrac{1}{2}(2A_2{}^2+A_3{}^3),\tag{14b}$$

$$t_V:\quad \tilde{Q}=-A_1{}^1,\qquad \tilde{Y}=A_3{}^3,\qquad \tilde{I}_3=-\tfrac{1}{2}(2A_1{}^1+A_3{}^3).\tag{14c}$$

Now the first one, t_I, for which

$$\tilde{Y}=-A_1{}^1-A_2{}^2,\tag{15}$$

$$\tilde{I}_3=\tfrac{1}{2}(A_2{}^2-A_1{}^1)=\tfrac{1}{2}(B_2{}^2-B_1{}^1)=I_3\tag{16}$$

corresponds to the I triplet mentioned above.

The structure of the remaining triplets t_U and t_V can be brought to much more transparent and symmetric forms in terms of the U-spin and V-spin subalgebras.[10] As in the case of relations (9) and (10) for $SU(3)$ and $U(3)$, we define the U and V spin of $U(3)$ in exactly the same forms as in $SU(3)$ except that all quantities are tilded quantities. From the $SU(3)$ definitions,[12] we then have

$$\tilde{Y}_U=-\tilde{Q}=-A_2{}^2-A_3{}^3,\tag{17}$$

$$\tilde{U}_3=\tilde{Y}-\tfrac{1}{2}\tilde{Q}=\tfrac{1}{2}(A_3{}^3-A_2{}^2)=\tfrac{1}{2}(B_3{}^3-B_2{}^2)=U_3\tag{18}$$

for (14b), and

$$\tilde{Y}_V=\tilde{Q}-\tilde{Y}=-A_3{}^3-A_1{}^1,\tag{19}$$

$$\tilde{V}_3=-\tfrac{1}{2}(\tilde{Y}+\tilde{Q})=\tfrac{1}{2}(A_1{}^1-A_3{}^3)=\tfrac{1}{2}(B_1{}^1-B_3{}^3)=V_3\tag{20}$$

for (14c). They correspond, therefore, to a U triplet and a V triplet, respectively, and hence the notations t_I, t_U, and t_V. With respect to the $SU(3)$ triplet (quark), these $U(3)$ triplets have their respective "hypercharges" (i.e., Y, Y_U, and Y_V) shifted by the amount of $\frac{2}{3}$ and as such they have quite different transformation properties under the Weyl reflections W_1, W_2, and W_3[11] which are reflections about the axis $I_3=0$, $U_3=0$, and $V_3=0$, respectively. Whereas the $SU(3)$ triplet is invariant under all three Weyl reflections, the $U(3)$ triplets are not. They transform according to

$$W_1:\quad t_I\to t_I,\quad t_U\leftrightarrow t_V;\tag{21a}$$

$$W_2:\quad t_U\to t_U,\quad t_I\leftrightarrow t_V;\tag{21b}$$

$$W_3:\quad t_V\to t_V,\quad t_I\leftrightarrow t_U.\tag{21c}$$

Figure 1 and Table I(a) list the quantum numbers $\tilde{I}_3$ and $\tilde{Y}$ for the single triplet (quark) model; a possible assignment implied by the two-triplet model[2] is shown in Fig. 2 and Table I(b); the corresponding quantum numbers for the three-triplet model are given in Fig. 3 and Table I(c).

TABLE I. Quantum-number assignments for (a) the quark model, (b) the two-triplet model, and (c) the three-triplet model.

(a)									
		quark							
$\tilde{I}_3$	$\frac{1}{2}$	$-\frac{1}{2}$	0						
$\tilde{Y}$	$\frac{1}{3}$	$\frac{1}{3}$	$-\frac{2}{3}$						
$\tilde{Q}$	$\frac{2}{3}$	$-\frac{1}{3}$	$-\frac{1}{3}$						
(b)									
		t_1			t_2				
$\tilde{I}_3$	$\frac{1}{2}$	$-\frac{1}{2}$	0	$\frac{1}{2}$	$-\frac{1}{2}$	0			
$\tilde{Y}$	1	1	0	-1	-1	-2			
$\tilde{Q}$	1	0	0	0	-1	-1			
(c)									
		$t_1(t_I)$			$t_2(t_U)$			$t_3(t_V)$	
$\tilde{I}_3$	$\frac{1}{2}$	$-\frac{1}{2}$	0	0	-1	$-\frac{1}{2}$	1	0	$\frac{1}{2}$
$\tilde{Y}$	1	1	0	0	0	-1	0	0	-1
$\tilde{Q}$	1	0	0	0	-1	-1	1	0	0

III. DOUBLE $SU(3)$ SYMMETRY

Let us call the three triplets $t_1(=t_I)$, $t_2(=t_U)$, and $t_3(=t_V)$. Each triplet may be characterized in general by the average values, $\bar{I}_3$ and $\bar{Y}$, of $\tilde{I}_3$ and $\tilde{Y}$ for its three members. This specifies the location of the center of the triplet in the $\tilde{I}_3-\tilde{Y}$ diagram. Since $\bar{A}_1{}^1=\bar{A}_2{}^2=\bar{A}_3{}^3=\bar{\tau}/3=\tau/3$, Eq. (14) gives for the three definitions of

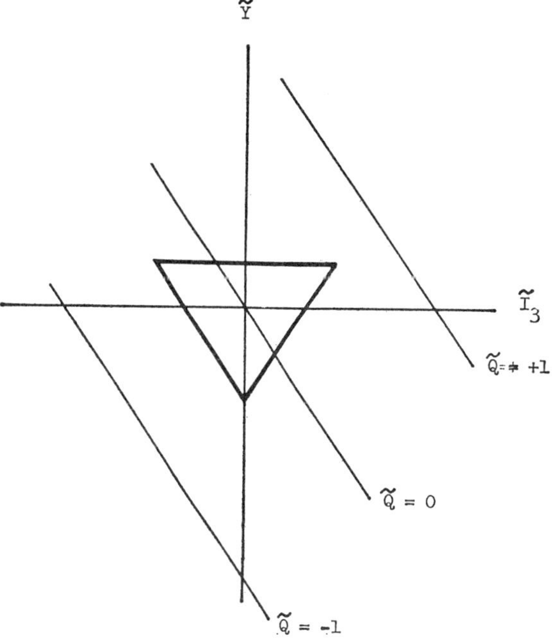

FIG. 1. The single-triplet (quark) model.

B 1009 THREE-TRIPLET MODEL WITH DOUBLE $SU(3)$ SYMMETRY

$\bar{I}_3$ and $\bar{Y}$,

$$\bar{I}_3=0,\ \tfrac{1}{2}\tau,\ -\tfrac{1}{2}\tau,\qquad \bar{Y}=-\tfrac{2}{3}\tau,\ \tfrac{1}{3}\tau,\ \tfrac{1}{3}\tau, \tag{22}$$

respectively, where $\tau=-1$ for all the triplets. We may define new quantities I_3, Y and $Q=I_3+\frac{1}{2}Y$ by the relations:

$$\tilde{I}_3=\bar{I}_3+I_3,\quad \tilde{Y}=\bar{Y}+Y,\quad \tilde{Q}=\bar{I}_3+\tfrac{1}{2}\bar{Y}+I_3+\tfrac{1}{2}Y=\bar{Q}+Q. \tag{23}$$

It is clear that I_3 and Y play the role of $SU(3)$ generators within each triplet. The charm number C defined in the two-triplet model[2] is then

$$\tfrac{1}{3}C=\bar{Q}=\bar{I}_3+\tfrac{1}{2}\bar{Y}. \tag{24}$$

Now it is interesting to note that according to Eq. (22) and Fig. 3, the centers of the three triplets form an antitriplet, equivalent to an antiquark, symmetrically located around the origin. Let us suppose that the nine members of the three triplets $t_{1\alpha}$, $t_{2\alpha}$, $t_{3\alpha}$, $\alpha=1, 2, 3$ be combined into a single multiplet $T=\{t_{i\alpha}\}$, $i=1, 2, 3$. We can then imagine two distinct sets of $SU(3)$ operations on T. One is the $SU(3)$ acting on the index α for each triplet, while the other $SU(3)$ acts on the index i, which mixes corresponding members of different triplets. T is then a representation $(3,3^*)$ of this group $G\equiv SU(3)'\times SU(3)''$.[14] The quantum numbers of $SU(3)'$ and

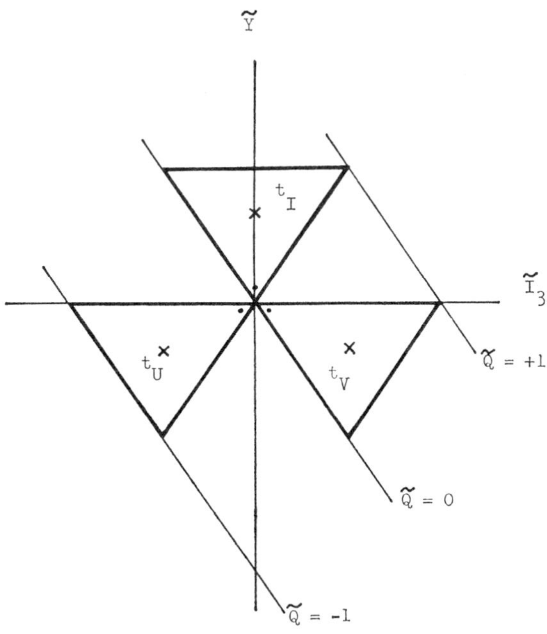

Fig. 3. The three-triplet model.

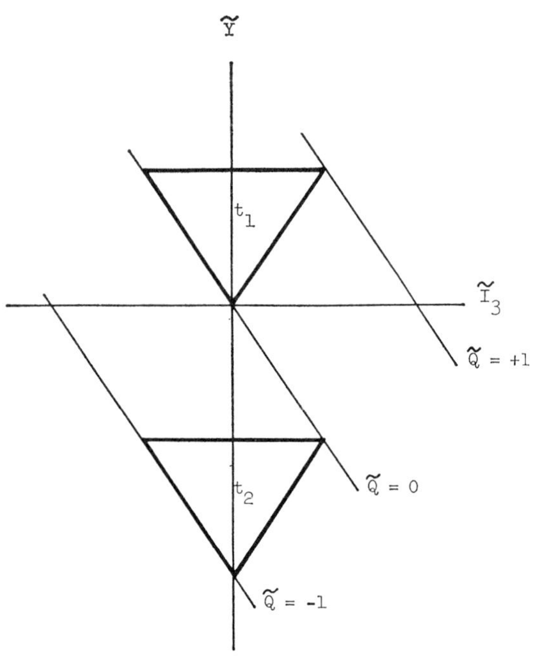

Fig. 2. The two-triplet model.

$SU(3)''$ are identified as $I_3'=I_3$, $Y'=Y$, $I_3''=\bar{I}_3$ and $Y''=\bar{Y}$ in Eq. (22), so that

$$\tilde{I}_3=I_3'+I_3'',\quad \tilde{Y}=Y'+Y'',\quad \tilde{Q}=I_3'+I_3''+\tfrac{1}{2}Y'+\tfrac{1}{2}Y'',\quad \tfrac{1}{3}C=I_3''+\tfrac{1}{2}Y''. \tag{25}$$

A general representation of G may be characterized by four numbers p', q', p'', q'' so that $D(p',q',p'',q'')\sim D(p',q')\times D(p'',q'')$, where $D(p,q)$ is a representation of $SU(3)$. However, in our scheme where the nonet T is the fundamental field, we do not get all the possible representations of G. This can be illustrated by means of the triality numbers[5] $t'=p'-q' \bmod(3)$, $t''=p''-q'' \bmod(3)$. The nonet T has $t'=1$, $t''=-1$. All representations constructed out of T and T^* then satisfy $t'=-t''$.

Let us next consider the meson and baryon states $\sim TT^*$ and $\sim TTT$. The $SU(3)'\times SU(3)''$ contents of these 81- and 729-plets are

$$\begin{aligned}(3,3^*)\times(3^*,3)&=(8,1)+(1,1)+(1,8)+(8,8),\\ (3,3^*)\times(3,3^*)\times(3,3^*)&=(1,1)+2(8,1)+2(1,8)\\ &\quad+(1,10^*)+(10,1)+2(8,10^*)+2(10,8)\\ &\quad+4(8,8)+(10,10^*).\end{aligned} \tag{26}$$

It is an attractive possibility to postulate at this point that the energy levels are classified according to $SU(3)''$. The masses will then depend on the Casimir operators of $SU(3)''$. For example, a simple linear form will be

$$m=m_0+m_2C_2''+m_3C_3'', \tag{27}$$

[14] Such a nonet provides a natural basis for the symmetry of $SU(9)$. However, we will not consider it here.

where C_2'', C_3'' are the eigenvalues of quadratic and cubic Casimir operators of $SU(3)''$. In particular, we may assume that the main mass splitting comes from C_2''. Since this increases with the dimensionality of representation, the lowest mass levels will be $SU(3)''$ singlets. This selects the low-lying meson and baryon states to be (8,1), (1,1) and (8,1), (1,1), (10,1), respectively. In general, all low-lying states will have triality zero, $t'=t''=0$.

As for the baryon number assignment to the triplets, the simplest possibility would be to assign an equal baryon number, i.e., $B=\frac{1}{3}$, to them. In this case the triplets themselves would be essentially stable, and their nine members would behave like an octet plus a singlet of "heavy baryons" as may be seen from Fig. 3. Another simple possibility may be $B=\frac{1}{3}+Y''$, namely $B=(1,0,0)$ for (t_1,t_2,t_3). We expect a mass splitting depending on B or Y'', which may be the origin of the Okubo-Gell-Mann mass formula.

The advantage of the three-triplet model is that the $SU(6)$ symmetry can be easily realized with s-state triplets. The extended symmetry group becomes now $SU(6)'\times SU(3)''$. Since an $SU(3)''$ singlet is antisymmetric, the over-all Pauli principle requires the baryon states to be the symmetric $SU(6)$ 56-plet. Other $SU(6)$ representations such as the 70, will be obtained by bringing in either the orbital angular momentum or the "ρ spin" of the Dirac spinor triplets.

As in the two-triplet model mentioned in the Introduction, the mass formula of the type (27) may be derived dynamically. Instead of the charm number field, we introduce now eight gauge vector fields which behave as (1,8), namely as an octet in $SU(3)''$, but as singlets in $SU(3)'$. Since their coupling to the individual triplets is proportional to λ_i'' [the generators of $SU(3)''$], the interaction energy arising from the exchange of these vector fields will yield the first and second terms of Eq. (27). If these mesons obey again a similar type of mass formula, they will be expected to be massive compared to the ordinary mesons. However, it is not clear whether the resulting short-range character of the interaction can be readily reconciled with the postulated largeness of the interaction energy.

We may characterize the hierarchy of interactions and their symmetries implied by the above model as follows. First, the *superstrong* interactions responsible for forming baryons and mesons have the symmetry $SU(3)''$, and causes large mass splittings between different representations. The scale of mass involved would be comparable or large compared to the baryon mass, namely $\gtrsim 1$ BeV. The lowest states, i.e., $SU(3)''$ singlet states, would split according to $SU(3)'$, which would be the $SU(3)$ group observed among the known baryons and mesons, with their *strong* interactions. The scale of mass splitting would then be $\lesssim 1$ BeV.

When we go to the massive $SU(3)''$ nonsinglet states, there may very well be coupling between the two $SU(3)$ groups similar to the $L\cdot S$ coupling. The levels should be classified in terms of the three sets of Casimir operators formed out of λ_i', λ_i'', and $\lambda_i=\lambda_i'+\lambda_i''$, respectively. The splitting due to the coupling would naturally be intermediate between the above two splittings, namely ~ 1 BeV. Because of this coupling, the separate conservation of the two $SU(3)$ spins, I_3' and Y' on the one hand, and I_3'' and Y'' on the other, would be destroyed, and only the sums $I_3=I_3'+I_3''$ and $Y=Y'+Y''$ would be conserved under *strong* interactions. This in turn would mean that all the massive states are in general highly unstable, and decay strongly to the low-lying states. (In the two-triplet model, we considered only weak decays of $C\neq 0$ states. But strong decays are also a possibility as is contemplated here.)

We have discussed here a possible model of baryons and mesons based on three triplets. How can we distinguish this and other different models mentioned already? Certainly different models predict considerably different structure of massive states. These states are characterized by the triality for the quark model, by the charm number for the two-triplet model and by the $SU(3)''$ representation for the present three-triplet model. If we restrict ourselves to the low-lying states only, however, it seems difficult to distinguish them without making more detailed dynamical assumptions.

ACKNOWLEDGMENTS

One of us (M. Y. H.) wishes to thank Professor E. C. G. Sudarshan and Professor A. J. Macfarlane for their encouragement and useful discussions and Professor L. O'Raifeartaigh and J. Kuriyan for helpful comments.

VOLUME 22, NUMBER 4 PHYSICAL REVIEW LETTERS 27 JANUARY 1969

HIGH-ENERGY ELECTROPRODUCTION AND THE CONSTITUTION OF THE ELECTRIC CURRENT*

C. G. Callan, Jr.,† and David J. Gross‡
Lyman Laboratory, Harvard University, Cambridge, Massachusetts
(Received 18 November 1968)

The asymptotic behavior of electroproduction cross sections is shown to contain information about the constitution of the electric current.

One of the most interesting possibilities which has emerged from the study of local current algebra is the use of high-energy inelastic lepton-nucleon scattering as a probe of the fundamental constituents of the electromagnetic current. In particular, Bjorken[1,2] has derived sum rules for equal-time commutators (ETCR) of space components of currents in terms of backward electron-scattering cross sections. Such commutators depend crucially on the constitution of the current in terms of fields, but the resulting sum rules are very difficult to evaluate, involving either neutrino scattering[1] or electroproduction on polarized targets.[2]

We would like to show that a much simpler test of this kind follows from the ETCR of the electromagnetic current with its time derivative. Depending on the constitution of the current, one finds that either longitudinal or transverse virtual photons dominate electroproduction cross sections for large momentum transfer. Consequently, simple cross-section measurements at high energies can yield fundamental information about the underlying theory.

Consider, first of all, the amplitude for virtual Compton scattering on protons,

$$T_{\mu\nu}(p,q)=i\int d^4x\, e^{iq\cdot x}\,\theta(x_0)\langle p|[J_\mu(x),J_\nu(0)]|p\rangle$$

$$=T_1(q^2,\nu)\left(-g_{\mu\nu}+\frac{q_\mu q_\nu}{q^2}\right)+T_2(q^2,\nu)\left(p_\mu-\frac{p\cdot q}{q^2}q_\mu\right)\left(p_\nu-\frac{p\cdot q}{q^2}q_\nu\right), \tag{1}$$

where $\nu=p\cdot q$, J_μ is the electromagnetic current, and subtraction of the disconnected part as well as an average over proton spins are understood. Both T_1 and T_2 satisfy fixed-q^2 dispersion relations (for spacelike q^2) which are conveniently written in terms of the variables $\omega=-q^2/\nu$:

$$T_1(q^2,\omega)=T_1(q^2,\infty)-\int_0^4 d\omega'^2\,\frac{W_1(\omega',q^2)}{\omega'^2-\omega^2+i\epsilon}, \tag{2a}$$

$$T_2(q^2,\omega)=-\omega^2\int_0^4\frac{d\omega'^2}{\omega'^2}\,\frac{W_2(\omega',q^2)}{\omega'^2-\omega^2+i\epsilon}. \tag{2b}$$

Following standard Regge lore, we have assumed that T_1 is once subtracted, while T_2 requires no subtraction. The quantity W_1 (W_2) is proportional to the differential cross section for backward (forward) inelastic lepton-proton scattering.

VOLUME 22, NUMBER 4 PHYSICAL REVIEW LETTERS 27 JANUARY 1969

Setting $q=(q_0,0)$, (1) becomes

$$T_{ij}=\delta_{ij}\left[T_1(q^2,\infty)-\int_0^4 d\omega'^2\frac{W_1(\omega',q^2)}{\omega'^2-q_0^2/p_0^2}\right]-\frac{p_ip_j}{m^2}\frac{q_0^2}{p_0^2}\int_0^4\frac{d\omega'^2}{\omega'^2}\frac{W_2(\omega',q^2)}{\omega'^2-q_0^2/p_0^2}. \tag{3}$$

Bjorken[3] has shown that the weak-interaction assumption

$$\lim_{q^2\to-\infty}(-q^2)\int_0^\infty\frac{d\nu}{\nu}W_2(q^2,\nu)<\infty \tag{4}$$

implies that both

$$F_1(\omega)=\lim_{\substack{q^2\to-\infty\\ \omega\text{ fixed}}} mW_1(\omega,q^2)$$

and

$$F_2(\omega)=\lim_{\substack{q^2\to-\infty\\ \omega\text{ fixed}}}\frac{\nu}{m}W_2(\omega,q^2) \tag{5}$$

exist. Therefore, if we multiply both sides of (3) by q_0^2 and let $q_0\to i\infty$, we obtain, with the help of Bjorken's asymptotic theorem,

$$C_{ij}(\bar{p})=-\int d^4x\,\delta(x_0)\langle p|[\dot{J}_i(x),J_j(0)]|p\rangle$$

$$=\delta_{ij}[\lim_{q^2\to-\infty}(q^2)T_1(q^2,\infty)]+\frac{p_0^2}{m}\int_0^4\frac{d\omega'^2}{\omega'}\left[\delta_{ij}\omega'F_1(\omega')-\frac{p_ip_j}{p_0^2}F_2(\omega')\right]. \tag{6}$$

Finally, we multiply this equation left and right by m/p_0^2 and take the limit $p_0\to\infty$, keeping the direction of $\vec{p}$ fixed:

$$\int_0^4\frac{d\omega'^2}{\omega'}\{\omega'F_1(\omega')(\delta_{ij}-\hat{p}_i\hat{p}_j)-[F_2(\omega')-\omega'F_1(\omega')]\hat{p}_i\hat{p}_j\}=\lim_{p_0\to\infty}\frac{m}{p_0^2}C_{ij}(\vec{p}). \tag{7}$$

The two quantities in the integrand have a simple expression in terms of the total cross sections for photoabsorption of transverse and longitudinal virtual photons:

$$\omega F_1(\omega)=\frac{1}{4\pi^2\alpha}\lim_{q^2\to-\infty}(q^2\sigma_T(\omega,q^2)), \tag{8}$$

$$F_2(\omega)-\omega F_1(\omega)=\frac{1}{4\pi^2\alpha}\lim_{q^2\to-\infty}(q^2\sigma_L(\omega,q^2)). \tag{9}$$

Clearly, if we know something about C_{ij}, (7) has consequences for the asymptotic behavior of $\sigma_{T,L}$.

It is instructive to compare two interesting and popular models: the algebra of fields[4] and the quark model with forces carried by a neutral vector meson (the so-called "gluon model"). The electromagnetic current is constituted differently in the two models, in the one case being proportional to a spin-1 field and in the other being bilinear in spin-$\frac{1}{2}$ fields. The ETCR of the current with its time derivative, and therefore C_{ij}, can be computed in both theories if one uses the equations of motion:

$$\delta(x_0)[\dot{J}_i(x),J_j(0)]=\delta(x)C_{ab}J_i^a(0)J_j^b(0)+c\text{-numbers (algebra of fields)}, \tag{10}$$

$$\delta(x_0)[\dot{J}_i(x),J_j(0)]=-\delta(x)\bar{\psi}\{i(\gamma_i\overleftrightarrow{\partial}_j+\gamma_j\overleftrightarrow{\partial}_i-2\vec{\gamma}\cdot\overleftrightarrow{\partial}\delta_{ij})-2g(\gamma_iB_j+\gamma_jB_i-2\vec{\gamma}\cdot\vec{B}\delta_{ij})+4M\delta_{ij}\}Q^2\psi$$

$$\text{(quark model)}, \tag{11}$$

VOLUME 22, NUMBER 4 PHYSICAL REVIEW LETTERS 27 JANUARY 1969

where J_μ^a are the SU(3) ⊗ SU(3) currents, C_{ab} is a numerical matrix, Q is the quark charge, and B_μ is the neutral vector-meson field. Just from the Lorentz tensor character of the commutator, one sees that in the two cases,[5]

$$C_{ij}(\vec{p}) = Ap_ip_j + A' \quad \text{(algebra of fields)},$$

$$C_{ij}(\vec{p}) = B(p_ip_j - \delta_{ij}\vec{p}^2) + B' \quad \text{(quark model)},$$

where A, A', B, B' are Lorentz scalars, about whose values we know nothing.

Upon applying this to (7) we get

$$\int_0^4 \frac{d\omega^2}{\omega^2}\{\omega F_1(\omega)(\delta_{ij} - \hat{p}_i\hat{p}_j) - [F_2(\omega) - \omega F_1(\omega)]\hat{p}_i\hat{p}_j\} = mA\hat{p}_i\hat{p}_j \quad \text{(algebra of fields)}$$
$$= mB(\hat{p}_i\hat{p}_j - \delta_{ij}) \quad \text{(quark model)}. \tag{12}$$

As Eqs. (8) and (9) indicate, ωF_1 and $F_2 - \omega F_1$ are positive quantities. Therefore, we must conclude that

$$4\pi^2\alpha\omega F_1(\omega) = \lim_{q^2 \to -\infty} q^2\sigma_T(\omega, q^2) = 0 \quad \text{(algebra of fields)}, \tag{13}$$

$$4\pi^2\alpha[F_2(\omega) - \omega F_1(\omega)] = \lim_{q^2 \to -\infty} q^2\sigma_L(\omega, q^2) = 0 \quad \text{(quark model)}. \tag{14}$$

It seems most likely, in fact, that the limit is approached as $1/q^2$, so that experimentally these two behaviors should be quite distinctive: Depending on which model of the current you choose σ_T/σ_L goes either as q^2 or $1/q^2$. The remarkable thing is that this result follows even though we have no information about the numerical value of the commutators (10), (11).[6]

The theorem also appears to be more general than the above model. In the quark model no matter what nonderivative interaction you choose, so long as the current has the general form $J_\mu = \bar{\psi}\gamma_\mu Q\psi$, it remains true that $q^2\sigma_L(\omega, q^2)$ vanishes for large spacelike q^2. In that sense the asymptotic vanishing of $q^2\sigma_L$ indicates that the electromagnetic current is made only out of spin-$\frac{1}{2}$ fields. If the current is bilinear in spin-0 fields one in general obtains the same theorem as for the algebra of fields. Not surprisingly, then, if the current contains both spin-0 and spin-$\frac{1}{2}$ fields, neither $q^2\sigma_T$ nor $q^2\sigma_L$ will vanish in the limit.[7]

We feel that these results are interesting on several counts. First of all, the connection between the asymptotic behavior of photoabsorption cross sections and the constitution of the current is as surprising as it is elegant. Second of all, the experimental verification of these theorems should be, in contrast to the previously known consequences of local current algebra, quite clean, since no sum rules have to be evaluated and no data on neutrons are needed. In fact, the rapid accumulation of data at the Stanford Linear Accelerator Center makes it likely that some aspects of these theorems can be compared with experiment in the near future.

---

*Research supported in part by the U. S. Office of Naval Research Contract No. Nonr-1866(55) (Mod. 1).
†Alfred P. Sloan Research Fellow.
‡Junior Fellow, Society of Fellows.

[1]J. D. Bjorken, Phys. Rev. 163, 1767 (1967). This sum rule can be rotated into an inequality for electroproduction on protons and neutrons. Unfortunately the integrals in these inequalities probably diverge logarithimically.

[2]J. D. Bjorken, Phys. Rev. 148, 1467 (1966).

[3]J. D. Bjorken, Stanford Linear Accelerator Center Report, 1968 (to be published). We would like to thank Professor Bjorken for informing us of his results prior to their publication.

[4]T. D. Lee, S. Weinberg, and B. Zumino, Phys. Rev. Letters 18, 1029 (1967).

[5]We are assuming that the singular nature of a product of local operators does not prevent us from deducing its tensor nature. In our applications the vacuum expectation value of the product is, of course, explicitly removed.

[6]In the algebra of fields, such commutators are explicitly equal to bilinear products of currents, and one might hope to relate their matrix elements to nonleptonic decay amplitudes. Unfortunately, on account of the subtraction in T_1 this does not work out.

Volume 22, Number 4 PHYSICAL REVIEW LETTERS 27 January 1969

[7]For completeness, we have also looked at the case of a canonical current constructed out of vector-meson fields. The arguments we have given must be modified since assumption (11) is violated. Nonetheless, one can conclude that σ_L/σ_T vanishes for large spacelike q^2.

Weak Interactions with Lepton-Hadron Symmetry*

S. L. Glashow, J. Iliopoulos, and L. Maiani†
Lyman Laboratory of Physics, Harvard University, Cambridge, Massachusetts 02139
(Received 5 March 1970)

We propose a model of weak interactions in which the currents are constructed out of four basic quark fields and interact with a charged massive vector boson. We show, to all orders in perturbation theory, that the leading divergences do not violate any strong-interaction symmetry and the next to the leading divergences respect all observed weak-interaction selection rules. The model features a remarkable symmetry between leptons and quarks. The extension of our model to a complete Yang-Mills theory is discussed.

INTRODUCTION

Weak-interaction phenomena are well described by a simple phenomenological model involving a single charged vector boson coupled to an appropriate current. Serious difficulties occur only when this model is considered as a quantum field theory, and is examined in other than lowest-order perturbation theory.[1] These troubles are of two kinds. First, the theory is too singular to be conventionally renormalized. Although our attention is not directed at this problem, the model of weak interactions we propose may readily be extended to a massive Yang-Mills model, which may be amenable to renormalization with modern techniques. The second problem concerns the selection rules and the relationships among coupling constants which are carefully and deliberately incorporated into the original phenomenological Lagrangian. Our principal concern is the fact that these properties are not necessarily maintained by higher-order weak interactions.

Weak-interaction processes, and their higher-order weak corrections, may be classified[2] according to their dependence upon a suitably introduced cutoff momentum Λ. Contributions to the S matrix of the form

$$\sum_{n=1}^{\infty} A_n(G\Lambda^2)^n$$

(where G is the usual Fermi coupling constant and A_n are dimensionless parameters) are called zeroth-order

* Work supported in part by the Office of Naval Research, under Contract No. N00014-67-A-0028, and the U. S. Air Force under Contract No. AF49(638)-1380.

† On leave of absence from the Laboratori di Fisica, Istituto Superiore di Santa, Roma, Italy.

[1] B. L. Ioffe and E. P. Shabalin, Yadern. Fiz. **6**, 828 (1967) [Soviet J. Nucl. Phys. **6**, 603 (1968)]; Z. Eksperim. i Teor. Fiz. Pis'ma v Redaktsiyu **6**, 978 (1967) [Soviet Phys. JETP Letters **6**, 390 (1967)]; R. N. Mohapatra, J. Subba Rao, and R. E. Marshak, Phys. Rev. Letters **20**, 1081 (1968); Phys. Rev. **171**, 1502 (1968); F. E. Low, Comments Nucl. Particle Phys. **2**, 33 (1968); R. N. Mohapatra and P. Olesen, Phys. Rev. **179**, 1917 (1969).

[2] T. D. Lee, Nuovo Cimento **59A**, 579 (1969).

weak effects, terms of the form

$$G \sum_{n=0}^{\infty} B_n(G\Lambda^2)^n$$

are called first-order weak effects, and generally, terms of the form

$$G^l \sum_{n=0}^{\infty} C_{ln}(G\Lambda^2)^n$$

are called lth order. (We are disregarding possible logarithmic dependences on the cutoff.) The zeroth-order terms present us with the dangerous possibility of serious violations of parity and hypercharge in strong interactions. First-order terms include the usual lowest-order contributions (order G) to leptonic and semileptonic processes. However, other first-order terms may yield violations of observed selection rules: There can be $\Delta S=2$ amplitudes, yielding a K_1-K_2 mass splitting, beginning at order $G(G\Lambda^2)$, as well as contributions to such unobserved decay modes as $K_2 \to \mu^+ + \mu^-$, $K^+ \to \pi^+ + l + \bar{l}$, etc., involving neutral lepton pairs, or departures from the leptonic $\Delta S = \Delta Q$ law. We shall say of a model that its divergences are properly ordered if it is true that the zeroth-order terms *do not* yield violations of parity or hypercharge, and if the first-order terms *do* satisfy the observed selection rules of weak-interaction phenomena.

In most conventional formulations of a weak-interaction field theory (say, a vector boson coupled to a quark triplet), the divergences are not properly ordered. Defenders of such theories must argue that there is an effective weak-interaction cutoff which guarantees that the induced higher-order effects are as small as experiment indicates. A remarkably small cutoff,[1] not greater than 3 or 4 GeV, seems necessary. Should such a cutoff be justified, the problem of higher-order departures from known selection rules is solved; all such departures are small.

Others feel that such a small cutoff is implausible and unrealistic, and that one must confront the possibility that $G\Lambda^2$ is large—perhaps obtaining sensible results in the limit $G\Lambda^2 \to \infty$. In this case, one may regard all the first-order terms as having the same general magnitude, that of observed weak phenomena, and nth-order terms as having the magnitude naively expected of nth-order weak interactions.

An elegant solution to the problem of the zeroth-order terms was recently discovered, removing the specter of strong violations of parity and hypercharge.[3,4] One assumes a particular form for the breakdown of chiral $SU(3)$: The symmetry-breaking term must transform like the $(3,\bar{3})+(\bar{3},3)$ representation[5]; in a quark model, like the quark mass term. In this case, the zeroth-order weak interactions may be identified as an object belonging to the same representation as the symmetry-breaking term. After an appropriate $SU(3) \times SU(3)$ transformation, their only effect is to cause a renormalization of the symmetry-breaking terms, giving renormalized quark masses.[4] There is no violation of hypercharge or parity. Indeed, from a speculative stability requirement of the symmetry-breaking term under weak and electromagnetic corrections, the correct value of the Cabibbo angle may be deduced.[4]

Although the zeroth-order terms are controlled with an appropriate model of strong interactions, the first-order terms remain troublesome. Indeed, with a quark model, we immediately encounter strangeness-violating couplings of neutral lepton currents and contributions to the neutral kaon mass splitting to order $G(G\Lambda^2)$.[6] (In such a model, departures from $\Delta S = \Delta Q$ first appear at second order.) For this reason, it appears necessary to depart from the original phenomenological model of weak interactions. One suggestion[7] involves the introduction of a large number of intermediaries of spins one and zero, so coupled that the leading divergences are associated with only the diagonal symmetry-preserving interactions; in this fashion a proper ordering of divergences is readily obtained. But this model is an awkward one involving many intermediaries with different spins but degenerate coupling strengths. Few would concede so much sacrifice of elegance to expediency.[8]

We wish to propose a simple model in which the divergences are properly ordered. Our model is founded in a quark model, but one involving four, not three, fundamental fermions; the weak interactions are mediated by just one charged vector boson. The weak hadronic current is constructed in precise analogy with the weak lepton current, thereby revealing suggestive lepton-quark symmetry. The extra quark is the simplest modification of the usual model leading to the proper ordering of divergences. Just as importantly, we argue that universality is preserved, in the sense that the

[3] C. Bouchiat, J. Iliopoulos, and J. Prentki, Nuovo Cimento **56A**, 1150 (1968); J. Iliopoulos, *ibid.* **62A**, 209 (1969); R. Gatto, G. Sartori, and M. Tonin, Phys. Letters **28B**, 128 (1968); Nuovo Cimento Letters **1**, 1 (1969).

[4] N. Cabibbo and L. Maiani, Phys. Letters **28B**, 131 (1968); Phys. Rev. D **1**, 707 (1970).

[5] S. L. Glashow and S. Weinberg, Phys. Rev. Letters **20**, 224 (1968); M. Gell-Mann, R. J. Oakes, and B. Renner, Phys. Rev. **175**, 2195 (1968).

[6] Of course, one cannot exclude *a priori* the possibility of a cancellation in the sum of the relevant perturbation expansion in the limit $\Lambda \to \infty$.

[7] M. Gell-Mann, M. L. Goldberger, N. M. Kroll, and F. E. Low, Phys. Rev. **179**, 1518 (1969).

[8] For other departures from the conventional theory, see, for example, C. Fronsdal, Phys. Rev. **136B**, 1190 (1964); W. Kummer and G. Segrè, Nucl. Phys. **64**, 585 (1965); G. Segrè, Phys. Rev. **181**, 1996 (1969); L. F. Li and G. Segrè, *ibid.* **186**, 1477 (1969); N. Christ, *ibid.* **176**, 2086 (1968). It should be understood that the ingenious conjecture of T. D. Lee and G. C. Wick [Nucl. Phys. **B9**, 209 (1969)] for removing divergences is logically independent of our analysis. If their hypothesis is correct, the role of the cutoff momentum is played by M_W. Only if M_W is small (~3-4 GeV) would the problems associated with ordering of divergences be solved; otherwise, a modification of the coupling scheme, such as ours, is still necessary.

leading divergent corrections (i.e., the first-order terms) yield a *common* renormalization to each of the various observed coupling constants.

The new model is discussed in Sec. I. Since Cabibbo's algebraic notion of universality[9] is maintained, that is to say, the entire weak charges generate the algebra of $SU(2)$, we observe in Sec. II that an extension to a three-component Yang-Mills model may be feasible. In contradistinction to the conventional (three-quark) model, the couplings of the neutral intermediary—now hypercharge conserving—cause no embarrassment. The possibility of a synthesis of weak and electromagnetic interactions is also discussed.

In Sec. III we briefly note some of the implications of the existence of a fourth quark, and finally, in Sec. IV we discuss some of the experimental tests of our model of weak interactions.

I. NEW MODEL

We begin by introducing four quark fields.[10] The three quarks $\mathcal{P}$, $\mathfrak{N}$, and λ form an $SU(3)$ triplet, and the fourth, $\mathcal{P}'$, has the same electric charge as $\mathcal{P}$ but differs from the triplet by one unit of a new quantum number $\mathcal{C}$ for charm. The strong-interaction Lagrangian is supposed to be invariant under chiral $SU(4)$, except for a symmetry-breaking term transforming, like the quark masses, according to the $(4,\bar{4})+(\bar{4},4)$ representation. This term may always be put in real diagonal form by a transformation of $SU(4)\times SU(4)$, so that B, Q, Y, $\mathcal{C}$, and parity are necessarily conserved by these strong interactions.

The extra quark completes the symmetry between quarks and the four leptons ν, ν', e^-, and μ^-. Both quadruplets possess unexplained unsymmetric mass spectra, and consist of two pairs separated by one in electric charge.

The weak lepton current may be expressed as

$$J_\mu{}^L=\bar{l}C_L\gamma_\mu(1+\gamma_5)l, \tag{1}$$

where l is a column vector consisting of the four lepton fields (ν, ν', e^-, μ^-) and the matrix C_L is given by

$$C_L=\begin{bmatrix}0&0&1&0\\0&0&0&1\\0&0&0&0\\0&0&0&0\end{bmatrix}. \tag{2}$$

This is a convenient way to rewrite the conventional current. In analogy with this expression, we define the weak hadron current to be

$$J_\mu{}^H=\bar{q}C_H\gamma_\mu(1+\gamma_5)q, \tag{3}$$

where q is the quark column vector $(\mathcal{P}',\mathcal{P},\mathfrak{N},\lambda)$ and the matrix C_H must be of the form

$$C_H=\left[\begin{array}{cc|c}0&0&\\0&0&U\\\hline 0&0&0\quad 0\\0&0&0\quad 0\end{array}\right] \tag{4}$$

in order for $J_\mu{}^H$ to carry unit charge. Pursuing the analogy further, we demand that the 2×2 submatrix U be unitary, so that the matrix C_H is equivalent to C_L under an $SU(4)$ rotation. Of course, it is not convenient to carry out the transformation making C_H and C_L coincide, for this would destroy the diagonalization of the $SU(4)$-breaking term, the quark masses. Nevertheless, suitable redefinitions of the relative phases of the quarks may be performed in order to make U real and orthogonal, so without loss of generality we write

$$U=\begin{bmatrix}-\sin\theta & \cos\theta\\ \cos\theta & \sin\theta\end{bmatrix}. \tag{5}$$

This is just the form of the weak current suggested in an earlier discussion of $SU(4)$ and quark-lepton symmetry.[10] What is new is the observation that this model is consistent with the phenomenological selection rules and with universality even when all divergent first-order terms [i.e., $G(G\Lambda^2)^n$] are considered.

To see this, we proceed diagrammatically in the quark model ignoring the strong $SU(4)$-invariant interactions.[11] Zeroth-order terms occur only in diagrams with only one external quark line, and give contributions to the quark mass operator of the form

$$\delta M(\gamma k)=\sum A_n(G\Lambda^2)^n\bar{q}M_n\gamma\cdot k(1+\gamma_5)q. \tag{6}$$

The A_n are dimensionless parameters, and the matrix M_n is a symmetric homogeneous polynomial of order n in C_H and of order n in $C_H{}^\dagger$. From the definition of C_H, it is seen that M_n must be a multiple of the unit matrix—again in contradistinction to the $SU(3)$ situation. Now, the zeroth-order terms are $SU(4)$ invariant.

There remains an apparent zeroth-order violation of parity, which may be transformed away because of the simple fashion of chiral $SU(4)$ breaking we have assumed. We simply define new quark fields

$$q_i'=(\alpha+\beta\gamma_5)q_i \tag{7}$$

with the real cutoff-dependent parameters α and β chosen so that the entire (bare plus zeroth-order) mass operator, in terms of q_i', is diagonal and parity conserving. The $SU(4)\times SU(4)$-invariant strong interactions are left unchanged. The procedure is analogous

[9] N. Cabibbo, Phys. Rev. Letters **10**, 531 (1963).

[10] B. J. Bjorken and S. L. Glashow, Phys. Letters **11**, 255 (1964).

[11] All our results about the zero- and first-order selection rules are trivially extended to the case of an $SU(4)$-invariant strong interaction which consists of a neutral vector boson coupled to quark number, the so-called "gluon" model. The only results of this paper which might be affected by such an interaction are the universality conditions in Eq. (9).

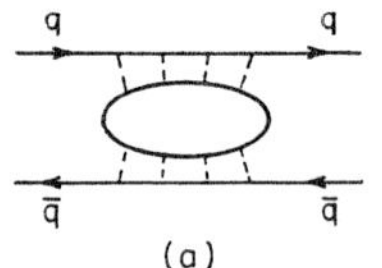

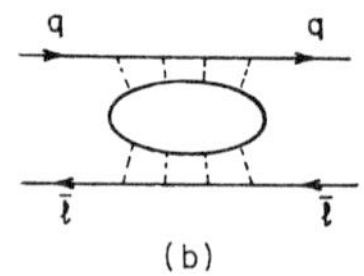

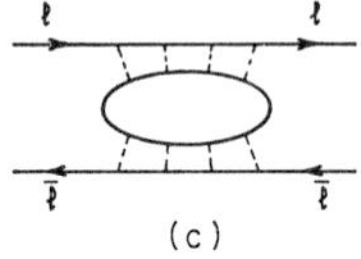

FIG. 1. (a) Connected part of the $q\bar{q}\to q\bar{q}$ amplitude. The crossed (annihilation) channel is also understood. (b) Connected part of the $ql\to ql$ amplitude. (c) Connected part of the $l\bar{l}\to l\bar{l}$ amplitude.

to that of Ref. 4, with the difference that the transformation (7) is $SU(4)$ invariant and does not change the definition of strangeness (or charm), or of the Cabibbo angle. An important consequence of the fact that M_n does not depend on the Cabibbo angle is that, unlike the situation in Ref. 4, it is impossible in our case to evaluate the Cabibbo angle by imposing a condition on the leading divergences. We conclude that zeroth-order weak effects are not significant.

We now consider the first-order $G(G\Lambda^2)^n$ terms which are of four types: (i) next-to-the-leading contributions to the quark and lepton mass operators, (ii) leading contributions to quark-quark or quark-antiquark scattering, (iii) leading contributions to quark-lepton scattering, and (iv) leading contributions to lepton-lepton scattering. Graphs with more than two external fermion lines yield no larger than second-order effects. Terms of type (i) are harmless: They contribute to observable nonleptonic $\Delta I=\frac{1}{2}$ processes, but since they cannot give $\Delta Y=2$, they do not produce a K_1K_2 mass splitting. On the other hand, type-(ii) diagrams could lead to $\mathfrak{N}\bar{\lambda}\to\bar{\mathfrak{N}}\lambda$, possibly giving rise to first-order contributions to the K_1K_2 mass difference, contrary to experiment. Let us show that they do not.

Graphs contributing to type (ii) effects are of the general form shown in Fig. 1(a), where the bubble includes any possible connections among the boson lines, and any number of closed fermion loops. The leading divergent contributions to q-$\bar{q}$ scattering from these graphs have the form

$$T_{HH}=G\sum_{n=2}^{\infty}B_n(G\Lambda^2)^{n-1}[\bar{q}\gamma_\mu(1+\gamma_5)\times B_H^{(n)}q\bar{q}\gamma^\mu(1+\gamma_5)B_H^{(n)\dagger}q], \quad (8)$$

where the B_n are finite dimensionless parameters independent of masses or momenta. It is clear that these first-order terms are independent of all external momenta. The matrix $B_H^{(n)}$ is a polynomial in C_H and $C_H^\dagger$ of order k and l, respectively, with $k+l\leq n$. Furthermore, the charge structure of the quark multiplets allows a change of charge no greater than unity, so that $|k-l|$ must be zero or one, and the matrices $B_H^{(n)}$ are easily computed (see the Appendix) to be

$$B_H^{(n)}=C_H \text{ or } C_H^\dagger \quad (k=l\pm1) \quad (8')$$
$$=[C_H,C_H^\dagger] \quad (k=l). \quad (8'')$$

Thus, T_{HH} gives rise to contributions with $|\Delta Y|\leq 1$ and, in particular, it does not yield a first-order K_1K_2 mass splitting. Of course, the next-to-the-leading divergences of these graphs will give $\Delta Y=2$, and do contribute to a second-order K_1K_2 mass difference, agreeing with experiment.

The leading divergences of types (iii) and (iv) give first-order contributions T_{HL} and T_{LL}, to semileptonic and leptonic processes. There will be a 1-to-1 correspondence among the graphs contributing to T_{LL}, T_{HL} [Figs. 1(b) and 1(c)], and T_{HH}. Because the algebraic properties of C_H and C_L are identical, we construct T_{HL} and T_{LL} from T_{HH} by the appropriate substitutions of $q\to L$ and $C_H\to C_L$.

In processes where the lepton charge changes, no violations of observed selection rules occur, but the first-order terms cause a renormalization of observed coupling constants. It is important to note that these renormalizations are common to leptonic and semileptonic processes, so that the relations

$$G_V(\Delta S=0)=G_\mu\cos\theta\,, \qquad G_V(\Delta S=1)=G_\mu\sin\theta \quad (9)$$

remain true when all first-order terms are included. This renormalization is given by the factor $1+\sum B_n(G\Lambda^2)^{n-1}$. A sufficient condition for these renormalizations to be common is the algebraic version of universality—a condition which is satisfied by our model, as well as by the usual three-quark model.

Next, we turn to the induced first-order couplings of hadrons to neutral lepton currents and self-couplings of neutral lepton currents. The neutral lepton currents are generated by the matrix C_L^0 and the neutral hadron currents by the matrix C_H^0, where

$$C_L^0=[C_L,C_L^\dagger]=\begin{bmatrix}1 & 0\\ 0 & -1\end{bmatrix}=[C_H,C_H^\dagger]=C_H^0. \quad (10)$$

Evidently, there are no induced couplings of neutral lepton currents to strangeness-changing currents. The induced couplings involve the strangeness-conserving current

$$\begin{aligned}J_\mu^0&=\bar{q}\gamma_\mu C_H^0(1+\gamma_5)q+\bar{l}\gamma_\mu C_L^0(1+\gamma_5)l\\ &=\bar{\mathcal{P}}'\gamma_\mu(1+\gamma_5)\mathcal{P}'+\bar{\mathcal{P}}\gamma_\mu(1+\gamma_5)\mathcal{P}-\bar{\mathfrak{N}}\gamma_\mu(1+\gamma_5)\mathfrak{N}\\ &\quad-\bar{\lambda}\gamma_\mu(1+\gamma_5)\lambda+\bar{\nu}'\gamma_\mu(1+\gamma_5)\nu'+\bar{\nu}\gamma_\mu(1+\gamma_5)\nu\\ &\quad-\bar{e}\gamma_\mu(1+\gamma_5)e-\bar{\mu}\gamma_\mu(1+\gamma_5)\mu\,. \quad (11)\end{aligned}$$

The coupling constant for this new neutral current-current interaction is a first-order expression of the form

$$G\sum_{n=2}^{\infty} C_n(G\Lambda^2)^{n-1}.$$

We anticipate that its strength should be comparable to the strength of the charged leptonic interactions. The new coupling plays no role in observed decay modes, but is should be detectable in accelerator experiments.

In Sec. II we discuss the possible extension of our model to a Yang-Mills model, where the coupling strength of the neutral W to its current is uniquely determined. These neutral lepton couplings constitute the most characteristic and interesting feature of our model. Relevant experimental evidence is discussed in Sec. IV.

II. YANG-MILLS MODEL OF WEAK INTERACTIONS

Divergences appear in our model of weak interactions, but they are properly ordered; observed selection rules are broken only in order $G^2(G\Lambda^2)^n$. But, the model is certainly not renormalizable. There is at least a possibility that a Yang-Mills model of weak interactions may be less singular.[12] In this section, we show how our model can be extended to include a symmetrically coupled triplet of W's. It is possible that W self-couplings can be introduced to give a complete Yang-Mills theory.

The Lagrangian with which we work may be written, in the four-quark model, without electromagnetism,

$$\mathfrak{L}=\mathfrak{L}_{\text{kin}}+\mathfrak{L}_s+\mathfrak{L}_M+\mathfrak{L}_w\,, \qquad (12)$$

where $\mathfrak{L}_{\text{kin}}$ is the purely kinematic term

$$\mathfrak{L}_{\text{kin}}=\bar{q}\gamma\cdot pq+\bar{l}\gamma\cdot pl+G_{\mu\nu}G^{\mu\nu}+W_{\mu\nu}{}^{\dagger}W^{\mu\nu} \qquad (13)$$

describing four free massless quarks, four leptons, and their strong and weak intermediaries ($X_{\mu\nu}$ denotes the antisymmetric curl of X_μ). $\mathfrak{L}_s$ denotes the $SU(4)$-invariant strong interaction, most simply

$$\mathfrak{L}_s=fG_\mu\bar{q}\gamma^\mu q\,, \qquad (14)$$

and $\mathfrak{L}_w$ is the weak interaction

$$\mathfrak{L}_w=gW_\mu{}^{\dagger}[\bar{q}C_H\gamma^\mu(1+\gamma_5)q+\bar{l}C_L\gamma^\mu(1+\gamma_5)l+\text{H.a.}]\,. \qquad (15)$$

The bare-mass term $\mathfrak{L}_M$ produces the observed masses of the leptons, the masses of W and G, and gives rise to the observed hierarchy of hadron symmetry,

$$\mathfrak{L}_M=\bar{q}M_Hq+\bar{l}M_Ll+m^2G_\mu G^\mu+M^2W_\mu W^\mu\,, \qquad (16)$$

where M_H and M_L are 4×4 matrices. This model gives a complete description of weak-interaction phenomena. The most important new feature is the appearance of neutral currents generated by the most divergent parts of diagrams containing an exchange of W^+, W^- pairs between two fermion lines. The effective coupling strength of these currents is expected to be of order G but, at this stage, we cannot predict its precise numerical value since we are unable to sum the perturbation series. In order to extend this model to a more symmetric one, we introduce an additional weak intermediary W_0 with appropriate couplings.

The couplings of W_0 to hadrons and leptons must be taken to be

$$2^{-1/2}gW_0{}^\mu\{\bar{q}[C_H{}^\dagger,C_H]\gamma_\mu(1+\gamma_5)q+\bar{l}[C_L{}^\dagger,C_L]\gamma_\mu(1+\gamma_5)l\}\,. \qquad (17)$$

We emphasize that the introduction of W_0 is by no means necessary in our model; however, we think that it gives a much more symmetric and aesthetically appealing theory.

In the conventional model of weak interactions, the extension to a three-component vector-meson theory cannot be made without contradicting experiment: The neutral boson leads to strangeness-changing decays involving neutral-lepton currents and to $\Delta S=2$ at order G. This is because the commutator of the conventional weak charge with its adjoint yields a strangeness-violating neutral charge. In our case, the corresponding operator is diagonal, and these difficulties are absent.

It is straightforward to show that the introduction of the neutral current does not spoil the proper ordering of divergences: The observed selection rules are preserved by all terms of order $G(G\Lambda^2)^n$. This is shown in the Appendix.

We note that W_0 is coupled to precisely the same neutral current appearing in the last section as an induced coupling. In the symmetric three-W model, its strength is uniquely predicted. Universality now applies to both charged and neutral couplings. That is to say, the leading divergent corrections to each are the same. The bare relationship

$$G_0=\tfrac{1}{2}G \qquad (18)$$

is preserved by the renormalizations, to first order [i.e., including all terms of order $G(G\Lambda^2)^n$]. This assertion is proved in the Appendix.

The introduction of a neutral W opens the possibility of formulating the weak interactions into a Yang-Mills theory. Self-couplings must be introduced among the W triplet in order to ensure the gauge symmetry. This is accomplished if we choose the Lagrangian in a manifestly gauge-invariant fashion:

$$\mathfrak{L}=\bar{q}\gamma\Pi_Hq+\bar{l}\gamma\Pi_Ll+\mathbf{W}_{\mu\nu}\mathbf{W}^{\mu\nu}+G_{\mu\nu}G^{\mu\nu}+\mathfrak{L}_M+\mathfrak{L}_s\,, \qquad (19)$$

where

$$\Pi_H{}^\mu=\partial^\mu+ig(\mathbf{C}_H\cdot\mathbf{W}^\mu)(1+\gamma_5)\,, \qquad (19')$$

$$\Pi_L{}^\mu=\partial^\mu+ig(\mathbf{C}_L\cdot\mathbf{W}^\mu)(1+\gamma_5)\,, \qquad (19'')$$

[12] See, for example, S. Mandelstam, Phys. Rev. **175**, 1580 (1969); M. Veltman, Nucl. Phys. **B7**, 637 (1968); H. Reiff and M. Veltman, *ibid.* **B13**, 545 (1969); D. Boulware, Ann. Phys. (N. Y.) **56**, 140 (1970); A. A. Slavnor, University of Kiev Report No. ITP 69/20 (unpublished); E. S. Fradkin and I. V. Tyutin, Phys. Letters **30B**, 562 (1969). Notice, however, that none of these references consider the far more difficult case of vector mesons coupled to nonconserved currents.

TABLE I. Quark quantum numbers.

	Fractional assignment			Integral assignment		
	Q	Y	$\mathcal{C}$	Q	Y	$\mathcal{C}$
$\mathcal{P}'$	$\frac{2}{3}$	$-\frac{2}{3}$	1	0	0	0
$\mathcal{P}$	$\frac{2}{3}$	$\frac{1}{3}$	0	0	0	-1
$\mathfrak{N}$	$-\frac{1}{3}$	$\frac{1}{3}$	0	-1	0	-1
λ	$-\frac{1}{3}$	$-\frac{2}{3}$	0	-1	-1	-1

and

$$W^{\mu\nu}=\Pi_W{}^\mu W^\nu-\Pi_W{}^\nu W^\mu\,, \qquad (19''')$$

where

$$(\Pi_W{}^\mu)_{ij}=\delta_{ij}\partial^\mu+ig2^{-1/2}(\mathbf{t}\cdot\mathbf{W}^\mu)_{ij}. \qquad (19'''')$$

The matrix-valued vectors $\mathbf{C}_H$ and $\mathbf{C}_L$ have components $(C,C^\dagger,2^{-1/2}[C^\dagger,C])$ in a basis where charge is diagonal, and $\mathbf{t}$ are the usual 3×3 generators of $O(3)$, with t_3 diagonal. The gauge group thus introduced is an exact symmetry of the entire Lagrangian excepting both $\mathcal{L}_M$ and electromagnetism.

The Yang-Mills model is undoubtedly the most attractive way to couple a triplet of vector mesons and the only one for which people have expressed some hope of constructing a renormalizable theory. The massless case has been proved to be renormalizable[12]; however, very little is known about the physically more interesting massive theory. In fact, the naive power counting shows that the highest divergence in a Yang-Mills theory is $g^{2n}\Lambda^N$ with $N=6n$. Notice that in the absence of the self-couplings the corresponding divergences are given, as we have already seen, by $N=2n$. So, at first sight, the Yang-Mills theory seems to be much more divergent than the ordinary coupling of the vector mesons with the currents. However, one can show that the naive limit $N=6n$ can be considerably lowered. We have already been able to show that $N\leq 3n$ and we believe that one can still lower this limit to at least $N=2n$. In other words, we believe that the introduction of the self-couplings does not make the theory more divergent.

Let us briefly consider a more daring speculation. It has long been suspected[13] that there may be a fundamental unity of weak and electromagnetic interactions, reflected phenomenologically by the common vectorial character of their couplings. For this reason, it may have been wrong for us to introduce a gauge symmetry for the weak interactions not shared by electromagnetism. As a more speculative alternative, consider the possibility of a four-parameter gauge group involving $\mathbf{W}$, and an additional Abelian singlet W_S, broken only by the mass term $\mathcal{L}_M$. Suppose, however, that a one-parameter gauge symmetry, corresponding to a linear combination A of W_0 and W_S remains unbroken. Then A must be massless, and may be identified as the photon. The orthogonal neutral combination B is massive, and acts as an intermediary of weak interactions along with $W^\pm$. This model could be correct only if the weak bosons are very massive (100 GeV) so that the weak and electromagnetic coupling constants could be comparable. With this model, the relation (18) would not persist, and the weak neutral current would involve $(1-\gamma_5)$ as well as $(1+\gamma_5)$ currents. The precise form of the model would depend on what linear combination of W_0 and W_S is the photon.

[13] J. Schwinger, Ann. Phys. (N. Y.) 2, 407 (1957); S. L. Glashow, Nucl. Phys. 10, 107 (1959); 22, 579 (1961)

III. ANOTHER QUARK MAKES $SU(4)$

Having introduced four quarks, we must consider strong interactions which admit the algebra of chiral $SU(4)$. Does this mean we should expect $SU(4)$ to be an approximate symmetry of nature? Nothing in our argument depends on how much $SU(4)$ is broken; the divergences are necessarily properly ordered. However, for the higher-order nonleading divergences to be as small as they must be, the breaking of $SU(4)$ cannot be too great: The limit on the cutoff Λ is replaced by a limit on Δ, a parameter measuring $SU(4)$ breaking; and from the observed K_1K_2 mass difference we now conclude that Δ must be not larger than 3–4 GeV. Thus, some residue of $SU(4)$ symmetry should persist.

We expect the appearance of charmed hadron states.[10] Meson multiplets, made up of a quark-antiquark pair, must belong to the 15-dimensional adjoint representation of $SU(4)$, consisting of an uncharmed $SU(3)$ singlet and octet, as well as two $SU(3)$ triplets of charm ±1. The structure of baryons depends on the quantum numbers assigned to the quarks. The two simplest possibilities are shown in Table I. For the more conventional fractional charge assignment, the baryons are made up of three quarks, and must belong to one of the representations contained in $4\times4\times4$. The only possibility is a 20-dimensional representation, which contains, besides the baryon octet, a triplet and sextet of charmed states and a doubly charmed triplet. The $j=\frac{3}{2}^+$ baryon decuplet belongs to another 20-dimensional representation with a charmed sectet, a doubly charmed triplet, and a triply charmed singlet.

With the integral-charge assignment, the baryon octet must be made of two quarks and an antiquark, the decuplet of three quarks and two antiquarks. The lepton and quark charged spectra now coincide, and the synthesis of weak and electromagnetic interactions appears more plausible. Moreover, there is no difficulty in obtaining the correct value for the π^0 lifetime.

Why have none of these charmed particles been seen? Suppose they are all relatively heavy, say $\sim$2 GeV. Although some of the states must be stable under strong (charm-conserving) interactions, these will decay rapidly ($\sim10^{+13}$ sec^{-1}) by weak interactions into a very wide variety of uncharmed final states (there are about a hundred distinct decay channels). Since the charmed particles are copiously produced only in associated production, such events will necessarily be of very complex topology, involving the plentiful decay products of both charmed states. Charmed particles could easily have escaped notice.

Finally, we briefly comment on the leptonic decay rates of ρ, ω, and ϕ (Γ_ρ, Γ_ω, and Γ_ϕ). Our electric current contains $SU(3)$ singlet as well as octet terms, so that the inequality

$$m_\omega\Gamma_\omega + m_\phi\Gamma_\phi \geq \tfrac{1}{3}m_\rho\Gamma_\rho \qquad (20)$$

may be deduced from the Weinberg spectral function sum rules and ω, ϕ, ρ dominance.[14] A stronger result is obtained if we extend Weinberg's Schwinger-term hypothesis to the vector currents of $SU(4)$:

$$m_\omega\Gamma_\omega + m_\phi\Gamma_\phi \geq m_\rho\Gamma_\rho. \qquad (21)$$

This result is in poor agreement with experiment, which favors the equality in (20). A resolution of this difficulty that does not abandon the Schwinger-term symmetry requires the introduction of a third $Y=T=0$ vector meson, another partner of ω and ϕ, corresponding to the $SU(4)$ singlet vector current.

IV. EXPERIMENTAL SUGGESTIONS

In this section, we discuss some of the observable effects characteristic of our picture of strong and weak interactions. Firstly, consider the experimental implications of the existence of a new quantum number—charm—broken only by weak interactions. The charmed particles, because they are heavy, are too short lived to give visible tracks. However, they should be copiously produced in hardonic collisions at accelerator energies:

$$(\text{hadron or } \gamma) + (\text{hadron}) \rightarrow X^{(+)} + X^{(-)} + \cdots,$$

where $X^{(\pm)}$ are oppositely charmed particles, each rapidly decaying into uncharmed hadrons with or without a charged lepton pair. The purely hadronic decay modes could provide illusory violations of hypercharge conservation in strong interactions. The leptonic decay modes provide a mechanism for the seemingly direct production of one or two charged leptons in hadron-hadron collisions.[15] Conceivably, muons thus produced may be responsible for the anomalous observed angular distribution of cosmic-ray muons in the 10^{12}-eV range,[16] where these directly produced muons may dominate the sea-level muon flux.

Should this last speculation about cosmic rays be correct, we need to revise radically estimates of the flux of ν and $\bar{\nu}$ in this energy range. We expect the charmed particle decays to yield equal numbers of each lepton variety; this gives a flux of electron neutrinos and antineutrinos equal to the muon flux, and 10–100 times greater than other estimates. This fact is of crucial importance to the possible detection of the resonance scattering[17]

$$\bar{\nu} + e^- \rightarrow \bar{\nu}' + \mu^-.$$

Charmed particles may be produced singly by neutrinos in such reactions as

$$\nu' + N \rightarrow \mu^- + X, \quad \bar{\nu}' + N \rightarrow \mu^+ + X,$$

where the charmed particle X would have a variety of decay modes, including leptonic ones. With the fractional charge assignment, the neutrino processes are suppressed by $\sin^2\theta$ and the antineutrino processes are forbidden. On the other hand, with the integral-charge assignment, the neutrino processes are again proportional to $\sin^2\theta$ while the antineutrino processes are proportional to $\cos^2\theta$.

The second new feature of our model is the appearance of neutral leptonic and semileptonic couplings involving a specified ($Y=0$) current and with a coupling constant comparable with the Fermi constant. Without the introduction of a W_0, we may say only $G_0 \sim G$. To be more definite, we shall phrase our arguments in terms of the value $G_0 = \frac{1}{2}G$ of Eq. (18).

Let us summarize the presently available data about these interactions.[18] Consider the following three reactions induced by muon neutrinos:

$$\text{(i)} \quad \nu' + e^- \rightarrow \nu' + e^-,$$

$$\text{(ii)} \quad \nu' + p \rightarrow \nu' + p,$$

$$\text{(iii)} \quad \nu' + p \rightarrow \nu' + \pi^+ + n.$$

None of these neutral couplings have been observed; experimentally, we can only quote limits. From the absence of observed forward energetic electrons in the CERN bubble-chamber experiments, we may conclude

$$G_0 \leq G,$$

a limit which is close to but consistent with our prediction.

For reaction (ii), it is found that

$$R = \sigma(\nu'p \rightarrow \nu'p)/\sigma(\bar{\nu}'p \rightarrow \mu^+ n) \leq 0.5$$

Because our neutral current contains both $I=0$ and $I=1$ parts, we cannot unambiguously predict this ratio. In a naive quark model, where the proton consists of only $\mathfrak{N}$ and $\mathcal{P}$ quarks, we find $R=\frac{1}{4}$, again close but consistent.

Finally, we quote the experimental limit on reaction (iii):

$$R' = \sigma(\nu' + p \rightarrow \pi^+ + n + \nu')/\sigma(\nu' + p \rightarrow \pi^+ + p + \mu^-) \leq 0.08.$$

[14] S. Weinberg, Phys. Rev. Letters **18**, 507 (1967); T. Das, V. Mathur, and S. Okubo, *ibid.* **19**, 470 (1967).

[15] In a recent experiment, P. J. Wanderer *et al.* [Phys. Rev. Letters **23**, 729 (1969)] have performed a search for W's by measuring the intensity and polarization of prompt energetic muons from the interaction of 28-GeV protons with nuclei. Their results are compatible with the assumption that all 25-GeV prompt muons have electromagnetic origin. There is no indication of the existence of W's. However, the published evidence does not seem to be relevant to the existence of charmed particles, which are produced in pairs, decay into many final states, and are not expected to produce many very energetic muons.

[16] H. E. Bergeson *et al.*, Phys. Rev. Letters **21**, 1089 (1968).

[17] M. G. K. Menon *et al.*, Proc. Roy. Soc. (London) **A301**, 137 (1967).

[18] See D. H. Perkins, in Proceedings of the Topical Conference in Weak Interactions, CERN, 1969 [CERN Report No. 69-7], pp. 1–42 (unpublished).

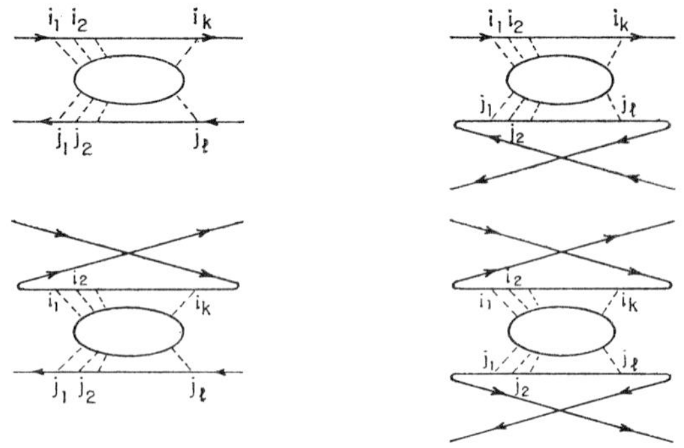

FIG. 2. Decomposition of the $q\bar{q} \to q\bar{q}$ connected amplitude by crossing the external fermion lines.

Because this transition is $\Delta I = 1$, we unambiguously predict $R' = \frac{1}{9}$ under the hypothesis of $\Delta(1238)$ dominance. In each of these three reactions, experiment is very close to a decisive test of our model.

In our model, the parity-violating nonleptonic interaction is also changed. In particular, the parity-violating one-pion-exchange nuclear force is no longer suppressed by $\sin^2\theta$.

Next we consider some experiments which could discover the existence of W_0. A simple and attractive possibility is the search for muon tridents in the semiweak reaction[19]

$$\mu^- + Z \to \mu^- + W_0 + Z,$$

with the subsequent muonic decay of W_0. Another possibility is the reaction[20]

$$e^+e^- \to \mu^+\mu^-.$$

The interference between the W^0 and photon contributions causes an asymmetry of the μ^+ angular distribution relative to the momentum of the incident e^+ given by

$$\delta = \frac{N_F - N_B}{N_F + N_B} = \frac{3M_W{}^2}{16\sqrt{2}} \frac{G}{\alpha\pi} \frac{s}{s - M_W{}^2},$$

where

$$G = 10^{-5} M_p{}^{-2}, \quad \alpha = 1/137, \quad \text{and} \quad s = 4E_e{}^2.$$

Away from the W^0 pole, the effect is rather small (less than 1% for $E_e = 3.5$ GeV) and it is masked by a similar effect due to the two-photon contribution. However, the factor $s/(s - M_W{}^2)$ makes the asymmetry increase sharply and change sign near M_W. Therefore, this reaction is an excellent tool to sweep a substantial mass range looking for W's. Another effect, much harder to detect, would be the direct observation of parity violation in $e^+e^- \to \mu^+\mu^-$. This requires the measurement of μ polarization.

Finally, we recall from Sec. III that the $SU(4)$ description of leptonic decays of vector mesons suggested the existence of another strongly coupled neutral $I = 0$ vector meson with considerable coupling to lepton pairs. Evidence for its existence could come from colliding beam experiments.

[19] M. Tannenbaum (private communication).

[20] N. Cabibbo and R. Gatto, Phys. Rev. 129, 1577 (1961).

APPENDIX

In this appendix we determine the form of the leading weak corrections to the q-$\bar{q}$, q-$\bar{l}$, and l-$\bar{l}$ amplitudes.

We have already shown that the wave-function renormalization of spinors is the same for both quarks and leptons and contributes a common factor to T_{HH}, T_{HL}, and T_{LL}. Therefore we need consider only the q-$\bar{q}$ amplitude T_{HH}. The other amplitudes T_{HL} and T_{LL} can be obtained from T_{HH} by appropriate substitutions. In the following, we shall omit the common wave-function renormalization factors.

For the sake of clarity, let us first consider our model of weak interactions, where we have three vector bosons symmetrically coupled.

The graphs of Fig. 1(a) can be decomposed into four classes of terms, obtained by keeping the boson lines fixed and reversing the fermion lines, as shown in Fig. 2. We then obtain for the contribution to T_{HH} corresponding to these four classes of diagrams

$$\begin{aligned} T_{HH}{}^{(n,k,l)} &= \bar{q}\gamma_\mu(1+\gamma_5)[C_{i_1}C_{i_2}\cdots C_{i_k} - (-1)^k C_{i_k}C_{i_{k-1}}\cdots C_{i_1}]q \\ &\quad \times P_{j_1\cdots j_l;\, i_1\cdots i_k}\bar{q}^\mu(1+\gamma_5) \\ &\qquad \times [C_{j_1}C_{j_2}\cdots C_{j_l} - (-1)^l C_{j_l}C_{j_{l-1}}\cdots C_{j_1}]q, \\ &\qquad\qquad k+l \leq n, \quad k, l \geq 1. \end{aligned} \tag{A1}$$

All the i's and j's go from 1 to 3 and the sum over all indices is understood. $P_{j_1\cdots j_l;\, i_1\cdots i_k}$ is a tensor made out of the invariant tensors δ_{ij} and ϵ_{ijk}.

It is easy to show that for any k

$$\mathrm{Tr}[C_{i_1}C_{i_2}\cdots C_{i_k} - (-)^k C_{i_k}C_{i_{k-1}}\cdots C_{i_1}] = 0. \tag{A2}$$

Therefore, since the interaction is $O(3)$ invariant, the connected part of T_{HH} has the form

$$T_{HH} = G \sum_{n=0}^{\infty} b_n (G\Lambda^2)^n \times (\bar{q}\gamma_\mu(1+\gamma_5)\mathbf{C}_H q)\cdot(\bar{q}\gamma^\mu(1+\gamma_5)\mathbf{C}_H q). \tag{A3}$$

In the case where we have only charged bosons, the argument is even simpler. Each of the indices $i_1\cdots i_k$, $j_1\cdots j_l$ appearing in Eq. (A1) takes only two possible values. With the relations

$$\begin{aligned} (C_H)^2 &= (C_H{}^\dagger)^2 = 0, \\ (C_H C_H{}^\dagger)^2 &= C_H C_H{}^\dagger, \\ (C_H{}^\dagger C_H)^2 &= C_H{}^\dagger C_H, \end{aligned} \tag{A4}$$

Eq. (A1) explicitly reads

$$T_{HH}{}^{(n;k,l)} = (\bar{q}\gamma_\mu(1+\gamma_5)[C_H, C_H{}^\dagger]q) \times (\bar{q}\gamma^\mu(1+\gamma_5)[C_H, C_H{}^\dagger]q), \quad k = l$$

$$T_{HH}{}^{(n,k,l)} = (\bar{q}\gamma_\mu(1+\gamma_5)C_H q)(\bar{q}\gamma^\mu(1+\gamma_5)C_H{}^\dagger q), \quad k = l+1. \quad \text{Q.E.D.}$$

Progress of Theoretical Physics, Vol. 49, No. 2, February 1973

CP-Violation in the Renormalizable Theory of Weak Interaction

Makoto KOBAYASHI and Toshihide MASKAWA

Department of Physics, Kyoto University, Kyoto

(Received September 1, 1972)

In a framework of the renormalizable theory of weak interaction, problems of *CP*-violation are studied. It is concluded that no realistic models of *CP*-violation exist in the quartet scheme without introducing any other new fields. Some possible models of *CP*-violation are also discussed.

When we apply the renormalizable theory of weak interaction[1] to the hadron system, we have some limitations on the hadron model. It is well known that there exists, in the case of the triplet model, a difficulty of the strangeness changing neutral current and that the quartet model is free from this difficulty. Furthermore, Maki and one of the present authors (T.M.) have shown[2] that, in the latter case, the strong interaction must be chiral $SU(4)\times SU(4)$ invariant as precisely as the conservation of the third component of the iso-spin I_3. In addition to these arguments, for the theory to be realistic, *CP*-violating interactions should be incorporated in a gauge invariant way. This requirement will impose further limitations on the hadron model and the *CP*-violating interaction itself. The purpose of the present paper is to investigate this problem. In the following, it will be shown that in the case of the above-mentioned quartet model, we cannot make a *CP*-violating interaction without introducing any other new fields when we require the following conditions: a) The mass of the fourth member of the quartet, which we will call ζ, is sufficiently large, b) the model should be consistent with our well-established knowledge of the semi-leptonic processes. After that some possible ways of bringing *CP*-violation into the theory will be discussed.

We consider the quartet model with a charge assignment of Q, $Q-1$, $Q-1$ and Q for p, n, λ and ζ, respectively, and we take the same underlying gauge group $SU_{\text{weak}}(2)\times SU(1)$ and the scalar doublet field φ as those of Weinberg's original model.[1] Then, hadronic parts of the Lagrangian can be devided in the following way:

$$\mathcal{L}_{\text{had}}=\mathcal{L}_{\text{kin}}+\mathcal{L}_{\text{mass}}+\mathcal{L}_{\text{strong}}+\mathcal{L}',$$

where $\mathcal{L}_{\text{kin}}$ is the gauge-invariant kinetic part of the quartet field, q, so that it contains interactions with the gauge fields. $\mathcal{L}_{\text{mass}}$ is a generalized mass term of q, which includes Yukawa couplings to φ since they contribute to the mass of q through the spontaneous breaking of gauge symmetry. $\mathcal{L}_{\text{strong}}$ is a strong-interaction part which conserves $\dot{I}_3$ and therefore chiral $SU(4)\times SU(4)$ invariant.[2] We assume C- and P-invariance of $\mathcal{L}_{\text{strong}}$. The last term denotes residual interaction parts if they exist. Since $\mathcal{L}_{\text{mass}}$ includes couplings with φ, it has possibilities of violating CP-conservation. As is known as Higgs phenomena,[3] three massless components of φ can be absorbed into the massive gauge fields and eliminated from the Lagrangian. Even after this has been done, both scalar and pseudoscalar parts remain in $\mathcal{L}_{\text{mass}}$. For the mass term, however, we can eliminate such pseudoscalar parts by applying an appropriate constant gauge transformation on q, which does not affect on $\mathcal{L}_{\text{strong}}$ due to gauge invariance.

Now we consider possible ways of assigning the quartet field to representations of the $SU_{\text{weak}}(2)$. Since this group is commutative with the Lorentz transformation, the left and right components of the quartet field, which are respectively defined as $q_L=\frac{1}{2}(1+\gamma_5)q$ and $q_R=\frac{1}{2}(1-\gamma_5)q$, do not mix each other under the gauge transformation. Then, *each* component has three possibilities:

$$A)\quad \mathbf{4}=\mathbf{2}+\mathbf{2}\,,$$

$$B)\quad \mathbf{4}=\mathbf{2}+\mathbf{1}+\mathbf{1}\,,$$

$$C)\quad \mathbf{4}=\mathbf{1}+\mathbf{1}+\mathbf{1}+\mathbf{1}\,,$$

where on the r.h.s., $\boldsymbol{n}$ denotes an n-dimensional representation of $SU(2)$. The present scheme of charge assignment of the quartet does not permit representations of $n\geq 3$. As a result, we have nine possibilities which we will denote by (A,A), (A,B), $\cdots$, where the former (latter) in the parentheses indicates the transformation properties of the left (right) component. Since all members of the quartet should take part in the weak interaction, and size of the strangeness changing neutral current is bounded experimentally to a very small value, the cases of (B,C), (C,B) and (C,C) should be abandoned. The models of (B,A) and (C,A) are equivalent to those of (A,B) and (A,C), respectively, except relative signs between vector and axial vector parts of the weak current. Since g_A/g_V ratios are measured only for composite states, this difference of the relative signs would be reduced to a dynamical problem of the composite system. So, we investigate in detail the cases of (A,A), (A,B), (A,C) and (B,B).

i) *Case* (A,C)

This is the most natural choice in the quartet model. Let us denote two $(SU_{\text{weak}}(2))$ doublets and four singlets by L_{d1}, L_{d2}, $R_{s1}^{(p)}$, $R_{s2}^{(p)}$, $R_{s1}^{(n)}$ and $R_{s2}^{(n)}$, where superscript $p(n)$ indicates p-like (n-like) charge states. In this case, $\mathcal{L}_{\text{mass}}$ takes, in general, the following form:

$$\mathcal{L}_{\text{mass}}=\sum_{i,j=1,2}[M_{ij}^{(n)}\bar{L}_{di}\varphi R_{sj}^{(n)}+M_{ij}^{(p)}\bar{L}_{di}\varepsilon\varphi^{*}R_{sj}^{(p)}]+\text{h.c.}\,,$$

$$\varphi^{*}=\begin{pmatrix}\varphi^{-}\\ \bar{\varphi}^{0}\end{pmatrix},\qquad \varepsilon=\begin{pmatrix}0 & 1\\ -1 & 0\end{pmatrix},\tag{1}$$

where $M_{ij}^{(n)}$ and $M_{ij}^{(p)}$ are arbitrary complex numbers. We can eliminate three Goldstone modes ϕ_i by putting

$$\varphi = e^{i\phi_i\tau_i}\begin{pmatrix} 0 \\ \lambda+\sigma \end{pmatrix}, \tag{2}$$

where λ is a vacuum expectation value of φ^0 and σ is a massive scalar field. Thereafter, performing a diagonalization of the remaining mass term, we obtain

$$\mathcal{L}_{\text{mass}} = \bar{q}mq\left(1+\frac{\sigma}{\lambda}\right),$$

$$m \equiv \begin{pmatrix} m_p & 0 & 0 & 0 \\ 0 & m_n & 0 & 0 \\ 0 & 0 & m_\zeta & 0 \\ 0 & 0 & 0 & m_\lambda \end{pmatrix}, \qquad q \equiv \begin{pmatrix} p \\ n \\ \zeta \\ \lambda \end{pmatrix}. \tag{3}$$

Then, the interaction with the gauge field in $\mathcal{L}_{\text{kin}}$ is expressed as

$$\sum_{j=1}^{3} A_\mu{}^j i\bar{q}\Lambda_j\gamma_\mu \frac{1+\gamma_5}{2} q\,. \tag{4}$$

Here, Λ_j is the representation matrix of $SU_{\text{weak}}(2)$ for this case and explicitly given by

$$\Lambda_+ = \frac{\Lambda_1+i\Lambda_2}{2} = K\begin{pmatrix} 0 & U \\ 0 & 0 \end{pmatrix}K^{-1}, \quad \Lambda_3 = \begin{pmatrix} 1 & 0 & 0 & 0 \\ 0 & 1 & 0 & 0 \\ 0 & 0 & -1 & 0 \\ 0 & 0 & 0 & -1 \end{pmatrix}, \quad K \equiv \begin{pmatrix} 1 & 0 & 0 & 0 \\ 0 & 0 & 1 & 0 \\ 0 & 1 & 0 & 0 \\ 0 & 0 & 0 & 1 \end{pmatrix}, \tag{5}$$

where U is a 2×2 unitary matrix. Here and hereafter we neglect the gauge field corresponding to $U(1)$ which is irrelevant to our discussion. With an appropriate phase convention of the quartet field we can take U as

$$U = \begin{pmatrix} \cos\theta & \sin\theta \\ -\sin\theta & \cos\theta \end{pmatrix}. \tag{6}$$

Therefore, if $\mathcal{L}'=0$, no CP-violations occur in this case. It should be noted, however, that this argument does not hold when we introduce one more fermion doublet with the same charge assignment. This is because all phases of elements of a 3×3 unitary matrix cannot be absorbed into the phase convention of six fields. This possibility of CP-violation will be discussed later on.

ii) *Case* (A, B)

This is a rather delicate case. We denote two left doublets, one right doublet and two singlets by L_{d1}, L_{d2}, R_d, $R_s{}^{(p)}$ and $R_s{}^{(n)}$, respectively. The general form of $\mathcal{L}_{\text{mass}}$ is given by

$$\mathcal{L}_{\text{mass}}=\sum_{i=1,2}[m_i\bar{L}_{di}R_d+M_i^{(n)}\bar{L}_{di}\varphi R_s^{(n)}+M_i^{(p)}\bar{L}_{di}\varepsilon\varphi^* R_s^{(p)}]+\text{h.c.},$$

where m_i, $M_i^{(n)}$ and $M_i^{(p)}$ are arbitrary complex numbers. After diagonalization of mass terms (in this case, the *CP*-odd part of coupling with σ does not disappear in general) each multiplet can be expressed as follows:

$$L_{d1}=\frac{1+\gamma_5}{2}\begin{pmatrix} p \\ \cos\theta e^{i\alpha}n+\sin\theta e^{i\beta}\lambda\end{pmatrix},\qquad L_{d2}=\frac{1+\gamma_5}{2}\begin{pmatrix} e^{i\gamma}\zeta \\ -\sin\theta e^{i\alpha}n+\cos\theta e^{i\beta}\lambda\end{pmatrix},$$

$$R_d=\frac{1-\gamma_5}{2}\begin{pmatrix}\sin\xi\cdot p+\cos\xi\cdot\zeta \\ \sin\eta\cdot n+\cos\eta\cdot\lambda\end{pmatrix},\qquad R_s^{(p)}=\frac{1-\gamma_5}{2}(\cos\xi\cdot p-\sin\xi\cdot\zeta),$$

$$R_s^{(n)}=\frac{1-\gamma_5}{2}(\cos\eta\cdot n-\sin\eta\cdot\lambda),\tag{7}$$

where phase factors α, β and γ satisfy two relations with the masses of the quartet:

$$e^{i\gamma}m_\zeta\sin\theta\cos\xi=m_p\cos\theta\sin\xi-e^{i\alpha}m_n\sin\eta\,,$$

$$e^{i\gamma}m_\zeta\cos\theta\cos\xi=-m_p\sin\theta\cos\xi+e^{i\beta}m_\lambda\cos\eta\,.\tag{8}$$

Owing to the presence of phase factors, there exists a possibility of *CP*-violation also through the weak current. However, the strangeness changing neutral current is proportional to $\sin\eta\cos\eta$ and its experimental upper bound is roughly

$$\sin\eta\cos\eta<10^{-2\sim-3}.\tag{9}$$

Thus, making an approximation of $\sin\eta\sim0$ (the other choice $\cos\eta\sim0$ is less critical) we obtain from Eq. (8)

$$m_\zeta/m_p\sim\cot\theta\cdot\tan\xi\,,$$

$$m_\lambda/m_n\sim\sin\xi/\sin\theta\,.\tag{10}$$

We have no low-lying particle with a quantum number corresponding to ζ, so that m_ζ, which is a measure of chiral $SU(4)\times SU(4)$ breaking, should be sufficiently large compared to the masses of the other members. However, the present experimental results on the g_A/g_V ratios of the octet baryon β-decay would not permit $\sin\xi>\sin\theta$. Thus, it seems difficult to reconcile the hierarchy of chiral symmetry breaking with the experimental knowledge of the semileptonic processes.

iii) *Case* (B, B)

As a previous one, in this case also, occurrence of *CP*-violation is possible, but in order to suppress $|\Delta S|=1$ neutral currents, coefficients of the axial-vector part of $\Delta S=0$ and $|\Delta S|=1$ weak currents must take signs oppossite to each other. This contradicts again the experiments on the baryon β-decay.

iv) *Case* (A, A)

In a similar way, we can show that no CP-violation occurs in this case as far as $\mathcal{L}'=0$. Furthermore this model would reduce to an exactly $U(4)$ symmetric one.

Summarizing the above results, we have no realistic models in the quartet scheme as far as $\mathcal{L}'=0$. Now we consider some examples of CP-violation through $\mathcal{L}'$. Hereafter we will consider only the case of (A, C). The first one is to introduce another scalar doublet field ϕ. Then, we may consider an interaction with this new field

$$\mathcal{L}'=\bar{q}\boldsymbol{\phi}\boldsymbol{C}\frac{1-\gamma_5}{2}q+\text{h.c.}, \tag{11}$$

$$\boldsymbol{\phi}=\begin{pmatrix} \bar{\phi}^0 & \phi^+ & 0 & 0 \\ -\phi^- & \phi^0 & 0 & 0 \\ 0 & 0 & \bar{\phi}^0 & \phi^+ \\ 0 & 0 & -\phi^- & \phi^0 \end{pmatrix}, \qquad \boldsymbol{C}=\begin{pmatrix} c_{11} & 0 & c_{12} & 0 \\ 0 & d_{11} & 0 & d_{12} \\ c_{21} & 0 & c_{22} & 0 \\ 0 & d_{21} & 0 & d_{22} \end{pmatrix},$$

where c_{ij} and d_{ij} are arbitrary complex numbers. Since we have already made use of the gauge transformation to get rid of the CP-odd part from the quartet mass term, there remains no such arbitrariness. Furthermore, we note that an arbitrariness of the phase of ϕ cannot absorb all the phases of c_{ij} and d_{ij}. So, this interaction can cause a CP-violation.

Another one is a possibility associated with the strong interaction. Let us consider a scalar (pseudoscalar) field S which mediates the strong interaction. For the interaction to be renormalizable and $SU_{\text{weak}}(2)$ invariant, it must belong to a $(4, 4^*)+(4^*, 4)$ representation of chiral $SU(4)\times SU(4)$ and interact with q through scalar and pseudoscalar couplings. It also interacts with φ and possible renormalizable forms are given as follows:

$$\begin{aligned} &\operatorname{tr}\{G_0S^+\boldsymbol{\varphi}\}+\text{h.c.}, \\ &\operatorname{tr}\{G_1S^+\boldsymbol{\varphi}G_2\boldsymbol{\varphi}^+S\}+\text{h.c.}, \\ &\operatorname{tr}\{G_1'S^+\boldsymbol{\varphi}G_2'S^+\boldsymbol{\varphi}\}+\text{h.c.}, \end{aligned} \tag{12}$$

with

$$\boldsymbol{\varphi}=\begin{pmatrix} \bar{\varphi}^0 & \varphi^+ & 0 & 0 \\ -\varphi^- & \varphi^0 & 0 & 0 \\ 0 & 0 & \bar{\varphi}^0 & \varphi^+ \\ 0 & 0 & -\varphi^- & \varphi^0 \end{pmatrix},$$

where G_i is a 4×4 complex matrix and we have used a 4×4 matrix representation for S. It is easy to see that these interaction terms can violate CP-conservation.

Next we consider a 6-plet model, another interesting model of *CP*-violation. Suppose that 6-plet with charges $(Q, Q, Q, Q-1, Q-1, Q-1)$ is decomposed into $SU_{\text{weak}}(2)$ multiplets as $2+2+2$ and $1+1+1+1+1+1$ for left and right components, respectively. Just as the case of (A, C), we have a similar expression for the charged weak current with a 3×3 instead of 2×2 unitary matrix in Eq. (5). As was pointed out, in this case we cannot absorb all phases of matrix elements into the phase convention and can take, for example, the following expression:

$$\begin{pmatrix} \cos\theta_1 & -\sin\theta_1\cos\theta_3 & -\sin\theta_1\sin\theta_3 \\ \sin\theta_1\cos\theta_2 & \cos\theta_1\cos\theta_2\cos\theta_3-\sin\theta_2\sin\theta_3 e^{i\delta} & \cos\theta_1\cos\theta_2\sin\theta_3+\sin\theta_2\cos\theta_3 e^{i\delta} \\ \sin\theta_1\sin\theta_2 & \cos\theta_1\sin\theta_2\cos\theta_3+\cos\theta_2\sin\theta_3 e^{i\delta} & \cos\theta_1\sin\theta_2\sin\theta_3-\cos\theta_2\sin\theta_3 e^{i\delta} \end{pmatrix} \tag{13}$$

Then, we have *CP*-violating effects through the interference among these different current components. An interesting feature of this model is that the *CP*-violating effects of lowest order appear only in $\Delta S\neq 0$ non-leptonic processes and in the semi-leptonic decay of neutral strange mesons (we are not concerned with higher states with the new quantum number) and not in the other semi-leptonic, $\Delta S=0$ non-leptonic and pure-leptonic processes.

So far we have considered only the straightforward extensions of the original Weinberg's model. However, other schemes of underlying gauge groups and/or scalar fields are possible. Georgi and Glashow's model[4] is one of them. We can easily see that *CP*-violation is incorporated into their model without introducing any other fields than (many) new fields which they have introduced already.

References

1) S. Weinberg, Phys. Rev. Letters **19** (1967), 1264; **27** (1971), 1688.
2) Z. Maki and T. Maskawa, RIFP-146 (preprint), April 1972.
3) P. W. Higgs, Phys. Letters **12** (1964), 132; **13** (1964), 508.
G. S. Guralnik, C. R. Hagen and T. W. Kibble, Phys. Rev. Letters **13** (1964), 585.
4) H. Georgi and S. L. Glashow, Phys. Rev. Letters **28** (1972), 1494.

Errata: The 3-3 component of the matrix in Eq (13) should read "$\cos\theta_1\sin\theta_2\sin\theta_3-\cos\theta_2\cos\theta_3 e^{i\delta}$"

VOLUME 13, NUMBER 5 PHYSICAL REVIEW LETTERS 3 AUGUST 1964

SPIN AND UNITARY SPIN INDEPENDENCE OF STRONG INTERACTIONS*

F. Gürsey† and L. A. Radicati‡
Brookhaven National Laboratory, Upton, New York
(Received 15 July 1964)

The purpose of this Letter is twofold. We want first to point out that the group SU(4) introduced by Wigner[1] to classify nuclear states can be extended to the relativistic domain and it is, therefore, relevant for particle physics. We will next show that when strangeness is taken into account the group SU(4) becomes enlarged to[2] SU(6) which contains, as a subgroup, $SU(3)\otimes[SU(2)]_q$. $[SU(2)]_q$ is the unitary subgroup (little group) of the Lorentz group that leaves invariant the momentum four-vector q.

The group we consider here embodies SU(3) and the ordinary spin in the same way as Wigner's SU(4) embodies isotopic spin and ordinary spin. Preliminary results on the classification of particles based on SU(6) seem encouraging enough to motivate a study of this group.[3]

We begin by discussing the first point. Let us assume that the ρ, ω, and π mesons are coupled to the nuclear field through a symmetrical Lagrangian of the form

$$L_{NM} = g\{\bar\psi\gamma_\mu\psi\omega_\mu + \bar\psi\gamma_\mu\tau^a\psi\rho_\mu^a + i\bar\psi\gamma_5\gamma_\mu\tau^a\psi\varphi_\mu^a\}, \quad (1)$$

where a denotes the isotopic spin index. Let us further impose the subsidiary conditions

$$\partial_\mu\omega_\mu = 0, \quad \partial_\mu\rho_\mu^a = 0,$$
$$\partial_\lambda\varphi_\mu^a - \partial_\mu\varphi_\lambda^a = 0, \quad (2)$$

which insure that $\omega_\mu, \rho_\mu^a, \varphi_\mu^a$ describe, respectively, particles with $(J=1^-, T=0)$, $(J=1^-, T=1)$, and $(J=0^-, T=1)$. The pion field π^a is related to the axial vector field φ_μ^a through $\varphi_\lambda^a = (1/\mu)\partial_\lambda\pi^a$, μ being the mass common to all mesons.

The conditions (2) are compatible with the equations of motion only if L includes, besides L_{NM} [Eq. (2)], additional terms such that the mesons (ρ, ω, π) are coupled to conserved currents. Thus ω and ρ are coupled to the conserved baryon and isotopic-spin currents, respectively, while the pion is coupled to a conserved axial-vector current.

It can now be shown[4] that L is invariant under a group[5] $\mathcal{G}_4$ which induces for each momentum q of the mesons a unitary unimodular transformation among the 15 degenerate states ω, ρ, and π. In counting the multiplicity we include, for a given momentum, the spin states just as for Wigner's supermultiplets. Under this transformation the nucleon $(S=\frac{1}{2}, T=\frac{1}{2})$ transforms like the four-dimensional representation of the group.

In the nonrelativistic limit, L_{NM} gives rise to a potential which describes spin- and isospin-independent exchange forces (Majorana forces) between nucleons. This potential is, therefore, invariant under Wigner's group SU(4). If now a purely spin-dependent perturbation is introduced, ω and ρ remain degenerate whereas the pion splits from them within the supermultiplet. We note that ω, ρ, and π are associated with the adjoint representation of SU(4). When this representation is reduced under the subgroup $SU(2)\otimes[SU(2)]_q$ it splits into states with $(J=1^-, T=0)$, $(J=1^-, T=1)$, and $(J=0^-, T=1)$.

These considerations are readily extended to include strange particles. In this case the SU(2)

VOLUME 13, NUMBER 5 PHYSICAL REVIEW LETTERS 3 AUGUST 1964

isotopic-spin group is replaced by SU(3) so that $\mathcal{G}_4$ goes over into a group $\mathcal{G}_6$ whose little group is $[SU(6)]_q$ which admits $SU(3)\otimes[SU(2)]_q$ as a subgroup.

The representation of SU(6) can be characterized by five integers $(\lambda_1\lambda_2\lambda_3\lambda_4\lambda_5)$ where the λ_i's are functions of the five Casimir operators. Table I shows some of the representations of SU(6) together with their SU(3) and spin structure. The symbols (m, n) in the third column refer to the SU(3) and spin multiplicity, respectively.

The lowest nontrivial representation (10000) has six dimensions. It represents a fundamental SU(3) triplet with (ordinary) spin $\frac{1}{2}$. Its SU(3) $\otimes$SU(2) content is (3, 2). The conjugate representation (00001) describes the antiparticle and its content is (3*, 2).

A Lagrangian similar to (1) can be written which couples invariantly the fundamental triplet to mesons corresponding to the 35-dimensional adjoint representation. When a spin-dependent perturbation is introduced the 35 states split into a pseudoscalar octet and a degenerate vector nonet[6] with negative parity. These can be identified with the observed (π, K, η) and $(\rho, \omega, K^*, \varphi)$ multiplets. SU(6) provides, therefore, a natural explanation of the degeneracy of the vector octet and the vector singlet in the nonet. All other meson-meson resonances must belong to self-conjugate representations of SU(6). Possible candidates are (00000) with even or odd parity, (10001) with even parity, (11011) with even or odd parity, etc.

The baryon octet and the $J=\frac{3}{2}^+$ decuplet can be grouped as a 56-dimensional representation obtained from the symmetrical combination of three fundamental triplets. The reduction of the direct product of $\underline{6}\otimes\underline{6}\otimes\underline{6}$ gives rise to three representations with 20, 70, and 56 dimensions. The fact that the ground state of the three-body configuration is symmetrical (56-dimensional representation) in the spin and unitary-spin variables implies that the two-body forces between them are repulsive. This seems to exclude a scheme based on only three fundamental quarks[7] whereas it is consistent with model II discussed in Appendix IV of reference 6. The connection of higher representations with possible baryon resonances is discussed by Pais.[3]

The splitting between the $J=0^-$ octet and the $J=1^-$ nonet suggests that the mass operator contains a spin-dependent term which can only be a function of $J(J+1)$. A simple mass formula for an SU(6) supermultiplet is the mass squared[6] formula

$$\mu^2=\mu_0^2+\alpha J(J+1)+\gamma[T(T+1)-\tfrac{1}{4}Y^2]$$

for mesons and

$$M=M_0+aJ(J+1)+bY+c[T(T+1)-\tfrac{1}{4}Y^2]$$

for baryons.

These are by no means the most general mass formulas that can be written on the basis of a broken SU(6) symmetry. The mass formula problem is further discussed by Pais.[3]

The interaction Lagrangian with conserved currents is generated from the free Lagrangian through a gauge transformation[4] associated with the group $\mathcal{G}_6$.[8] As in the case of the electromagnetic interaction this implies parity conservation for the strong interactions invariant under $\mathcal{G}_6$. Hence all the states of an SU(6) supermultiplet must have the same parity. Our scheme is, therefore, different from others that have been discussed recently[2,9,10]; in particular it does not predict 0^+ and 1^+ mesons degenerate with the existing 0^- and 1^- mesons. The degenerate states associated with the meson states for given momentum $\vec{q}$ and given SU(6) quantum numbers are simply the states corresponding to the opposite momentum and the same SU(6) quantum numbers.

It is a pleasure to thank G. C. Wick for very constructive criticism and A. Pais for stimulating discussions.

Table I. Some representations of SU(6) and their unitary spin and spin content.

Labeling $(\lambda_1\lambda_2\lambda_3\lambda_4\lambda_5)$	Dimensions $D(\lambda_1\lambda_2\lambda_3\lambda_4\lambda_5)$	Unitary spin and spin multiplicities (n, m)
(00000)	1	(1,1)
(10000)	6	(3,2)
(00001)	6*	(3*,2)
(01000)	15	(3*,3),(6,1)
(00100)	20	(8,2),(1,4)
(20000)	21	(6,3),(3*,1)
(10001)	35	(8,3),(8,1),(1,3)
(30000)	56	(10,4),(8,2)
(11000)	70	(10,2),(8,4),(8,2),(1,2)

*Work supported by the U. S. Atomic Energy Commission.

†On leave from the Middle East Technical University, Ankara, Turkey.

‡On leave from Scuola Normale Superiore, Pisa, Italy.

[1]E. P. Wigner, Phys. Rev. 51, 105 (1937). For recent evidence on the validity of the supermultiplet model,

VOLUME 13, NUMBER 5 PHYSICAL REVIEW LETTERS 3 AUGUST 1964

see P. Franzini and L. A. Radicati, Phys. Letters 6, 322 (1963).

[2]The group SU(6) has been suggested in a somewhat different context by M. Gell-Mann, to be published. Gell-Mann's point of view is, however, different from the one discussed here, being based on the algebra of the conserved and quasiconserved currents.

[3]For a more detailed analysis of the applications, see A. Pais, following Letter [Phys. Rev. Letters 13, (1964)].

[4]F. Gürsey and L. A. Radicati, to be published.

[5]The group $\mathcal{G}_4$ is noncompact and may be regarded as an extension of the Lorentz group by means of the isotopic spin group. The generators of $\mathcal{G}_4$ are the covariant spin operators, the isotopic spin operators and their products. The little group of $\mathcal{G}_4$ for fixed momentum q is $[SU(4)]_q$.

[6]F. Gürsey, T. D. Lee, and M. Nauenberg, Phys. Rev. 135, B467 (1964).

[7]M. Gell-Mann, Phys. Letters 8, 214 (1964).

[8]It is clear that the fundamental triplets will be coupled to the mesons through F-type coupling only. Since the baryons do not belong to the lowest representation of SU(6), the gauge operators generate a larger algebra which produces F-type couplings with the vector mesons and F- and D-type couplings with the pseudoscalar mesons.

[9]P. G. O. Freund and Y. Nambu, Phys. Rev. Letters 12, 714 (1964).

[10]A. Salam and J. C. Ward, unpublished.

Symmetric Quark Model of Baryon Resonances*

O. W. Greenberg† and M. Resnikoff‡
Department of Physics and Astronomy, University of Maryland, College Park, Maryland

(Received 9 June 1967)

The symmetric quark model of baryon resonances is applied to the (**56**,0^+), (**70**,1^-), and (**20**,1^+) supermultiplets. Using a systematic $SU(6)$ analysis, octet dominance, and dominance of two-body contributions to the mass operator, the Gürsey-Radicati mass formula for the (**56**,0^+) is derived without use of perturbation theory. An equal-spacing relation is derived for the $SU(6)$-symmetric mass contribution for the (**56**,0^+), (**70**,1^-), and (**20**,1^+) in ascending order, with the (**20**,1^+) lying above 2 BeV. A detailed analysis of the (**70**,1^-), which yields a quantitative fit for the spin-orbit-split negative-parity resonances, is made, with the result that the magnitude of the octet spin-orbit term is about six times greater than the singlet one. The masses and mixing amplitudes for all the resonances in the (**70**,1^-) are calculated, and it is pointed out that because of the large mixings for $J^P=\frac{1}{2}^-$ and $\frac{3}{2}^-$, the Gell-Mann–Okubo mass formula cannot be expected to hold for these resonances, so that there is no sense in trying to group them into octets and decuplets.

1. INTRODUCTION

THE introduction of quarks[1] or other triplets as fundamental constituents of hadrons has given a new unifying point of view to particle physics. This notion of fundamental triplets achieved striking success when combined with the idea of approximate-spin and unitary-spin symmetry of the interactions relevant to the low-lying hadronic states.[2] This $SU(6)$ theory[3] has led to a number of striking results which can be obtained in an elementary way using the quark model[4]: classification of baryon and meson supermultiplets,[2] mass formulas for these multiplets,[2] magnetic moment ratios for baryons,[5] and scattering of hadrons at high energy.[6] The underlying idea is that the observed hadrons are bound states of the quarks or other triplets. Up to now, a naive bound-state model in which the triplets are assumed to move nonrelativistically has been surprisingly successful, even though at present the model has no fundamental justification. One of the surprising features which has appeared from the study of the baryons is that the $SU(6)$ wave function of the baryons is totally symmetric.[2] Since ordinarily the ground state of a composite system has all particles in the lowest s state, the symmetry under permutations of the $SU(6)$ wave function indicates an apparent symmetric statistics for the triplets in the ground state of the baryons. Evidence against an antisymmetric space wave function for the ground state of the baryons comes from the study of the saturation of triplets bound in hadronic states[7] and from the consideration of the electromagnetic form factors of the proton and the neutron[8]; further, the form factors provide specific evidence for a symmetric ground-state space wave function, since the proton form factors and the neutron magnetic form factor remain positive in the range of momentum transfers in which they have been measured[8–10] (up to 245 F^{-2}).[11]

Two proposals have been put forward to explain the apparent violation of Fermi statistics in the ground state of the baryons. One of these is that the quarks are parafermions of order three,[12] in which case three quarks can be in a symmetric state under permutations and the composite object which is a bound state of three such quarks is a Fermi particle. This theory seems to be equivalent to a theory of three indistinguishable Fermi triplets. (One could also have a theory with three distinguishable fractionally charged Fermi triplets.) For the paraquark theory the triplets must have the usual quantum numbers associated with the quarks, for ex-

* This work was reported in an invited talk at the meeting of the American Physical Society in Washington, D. C. in April, 1967.

† Supported in part by the National Science Foundation under National Science Foundation Grant NSF GP 6036.

‡ Supported in part by the U. S. Air Force under contract AFOSR 500-66. Present address: Department of Physics, State University of New York at Buffalo.

[1] M. Gell-Mann, Phys. Letters **8**, 214 (1964); G. Zweig, CERN Reports 8182/TH.401 and 8419/TH.412, 1964, unpublished.

[2] F. Gürsey and L. A. Radicati, Phys. Rev. Letters **13**, 173 (1964); B. Sakita, Phys. Rev. **136**, B1756 (1964).

[3] A. Pais [Rev. Mod. Phys. **38**, 215 (1966)] and F. J. Dyson [*Symmetry Groups in Nuclear and Particle Physics* (W. A. Benjamin, Inc., New York, 1966)] contain reviews, reprints, and bibliographies.

[4] R. H. Dalitz [in *Proceedings of the Thirteenth International Conference on High-Energy Physics, Berkeley, California, 1966* (University of California Press, Berkeley, California, 1967), pp. 215–236] gave a survey of quark models in August, 1966.

[5] M. A. B. Bég, B. W. Lee, and A. Pais, Phys. Rev. Letters **13**, 514 (1964); B. Sakita, *ibid.* **13**, 643 (1964).

[6] E. M. Levin and L. L. Frankfurt, JETP Pis'ma v Redaktsiyu **2**, 105 (1965) [English transl.: JETP Letters **2**, 65 (1965)]; H. J. Lipkin and F. Scheck, Phys. Rev. Letters **16**, 71 (1966).

[7] O. W. Greenberg and D. Zwanziger, Phys. Rev. **150**, 1177 (1966).

[8] A. N. Mitra and R. Majumdar, Phys. Rev. **150**, 1194 (1966).

[9] R. E. Kreps and J. J. de Swart [University of Pittsburgh Report NYO-3829-4 (unpublished)] quote a theorem of L. K. Pandit and V. S. Mathur that the form factor must have at least one zero for any antisymmetric wave function. Kreps and de Swart argue that the zero in the form factor can be made to occur at arbitrarily large momentum transfer by proper choice of the antisymmetric wave function, and exhibit a 4-parameter antisymmetric wave function which leads to a form factor which fits the present data (see Ref. 10).

[10] S. Ishida, K. Konno, and H. Shimodaira [Nuovo Cimento **46A**, 189 (1966)] found form factors which fit the present data using a one-parameter symmetric wave function.

[11] W. Albrecht *et al.*, Phys. Rev. Letters **17**, 1192 (1966).

[12] O. W. Greenberg, Phys. Rev. Letters **13**, 598 (1964). The supermultiplets $(2L+1)$ dim$SU(6)$ are labelled [dim$SU(6)$, L^P] in the present article.

ample, fractional charge. The second proposal to explain the apparent violation of Fermi statistics is a three-triplet model[13] in which a new three-valued degree of freedom is introduced and there are nine fundamental particles. If one assumes that a new $SU(3)$ group called $SU(3)''$ acts on the new three-valued degree of freedom, then the lowest-lying baryons, those presently known, can be placed in the singlet of this $SU(3)''$ with an $SU(3)''$ wave function $\epsilon^{A''B''C''}$, A'', B'', $C''=1, 2, 3$. For this state each of the three $SU(3)''$ degrees of freedom occurs with equal probability and this degree of freedom is essentially averaged out. The quantum numbers of the triplets in these $SU(3)''$ singlet states are effectively replaced by their average values, which are those of the particles in the quark model. Thus this model has the attractive feature that the charges of the triplets are integral, but they have effectively fractional values in the observed baryons. [Similar remarks hold for the $SU(3)''$ singlet mesons.] The three-triplet model allows a construction of particles in the **10*** and other representations which cannot be obtained from the quark model using three-body states. Both the para-quark and three-triplet models introduce the number three explicitly as the maximum number of quarks[14] in a state which is symmetric under permutations. For most purposes in which the independent quarks do not appear singly these models are equivalent. We will refer to a model in which the three triplets in a baryon are symmetric under permutations as the *symmetric quark model*.

We abstracted this model from the properties of the ground state of the baryons. There are two possibilities to consider as candidates for the higher baryonic states: orbital excitation of quarks and $SU(6)$ excitation in which quark-antiquark pairs are added. It seems to be in keeping with the nonrelativistic picture used for the ground state to assume that quark-antiquark excitation is not important; in addition, it is attractive to use a model in which only three objects occur and to make use of the analogy of such a model to the systems studied in atomic and nuclear physics. In addition, with one exception, the Z_0^+ resonance which occurs in K^+N scattering,[15] there is no firm evidence for a resonance which does not fit into those $SU(3)$ multiplets which can be obtained from three quarks: namely, the singlet, octet, and decuplet. In this article we will study the orbital excitation model for the higher states.[12]

[13] Y. Nambu, in *Preludes in Theoretical Physics*, edited by A. de-Shalit, H. Feshbach, and L. Van Hove (North-Holland Publishing Company, Amsterdam, 1966), pp. 133–142; M. Y. Han and Y. Nambu, Phys. Rev. **139**, B1006 (1965); A. Tavkhelidze, in *Proceedings of the Seminar on High-Energy Physics and Elementary Particles, Trieste, 1965* (International Atomic Energy Agency, Vienna, 1965), pp. 763–779, and references cited therein.

[14] We will use the word "quark" for a particle which has an effectively fractional charge in the $SU(3)''$ singlet, as well as for a particle whose charge is really fractional.

[15] R. L. Cool *et al.*, Phys. Rev. Letters **17**, 102 (1966).

In order to restrict the number of *a priori* possible mass formulas for the supermultiplets of hadronic states, we will assume that the terms in the mass formulas come from only two sources: one-body effects which can be associated with the masses of the quarks, and two-body effects which can be associated with two-body interactions among the quarks. (Our actual analysis makes use of a parametrization of the mass of the observed resonances in terms of one- and two-body operators and might not require so concrete an identification of the terms.) Within this framework we make the most general analysis which is compatible with approximate $SU(6)$ symmetry and octet dominance. In particular, our derivation of mass formulas does not make use of perturbation theory.

The baryon resonances are of special interest for two reasons. First, the interesting question of symmetric statistics arises only for the baryons; the mesons look the same in the symmetric-quark model as in the Fermi quark model. Secondly, the simplifying assumption that one- and two-body effects dominate holds no advantage for the mesons which are two-body systems.

We carry out our analysis using artificial single-particle space states for the quarks. This simplifying assumption makes it easier to handle the symmetrization of the states and probably does not lead to serious error provided that certain pitfalls, such as excitation of center-of-mass motion (the "spurious states" of nuclear physics) are avoided. We think that this simplifying assumption is appropriate for a phenomenological analysis such as we are making.

For the higher states, the $SU(6)$ approximate symmetry which has been so successful for the ground state should be augmented by an L-S coupling scheme,[12] which sometimes goes by the name $SU(6)\times O(3)$. We have adopted this classification of the higher states and have also used the L-S coupling point of view in the analysis of the mass operator. It is convenient to refer to the supermultiplets in this model by expressions (dim $SU(6)$, L^P). Thus the ground state is $(\mathbf{56},0^+)$ and the lowest negative-parity baryon resonances are $(\mathbf{70},1^-)$. We are mainly interested in these negative-parity resonances in this article; however, we will also consider the $(\mathbf{20},1^+)$ because the parameters necessary to derive a mass formula for this supermultiplet are already completely determined by the $(\mathbf{56},0^+)$ and the $(\mathbf{70},1^-)$.

To analyze the one- and two-body operators, we make systematic use of the $SU(6)$ approximate symmetry. From this point of view we can find the most general one- and two-body operators, and classify them according to their spin dependence and according to irreducible representations of $SU(6)$. Experience in analyzing the masses of baryons and mesons indicates that the lowest $SU(6)$ representations are most important,[16] and we will make use of this fact in choosing

[16] H. Harari and M. A. Rashid, Phys. Rev. **143**, 1354 (1966).

which operators to retain for our calculation.[17] Having found the relevant operators, we will normalize them in such a way that the parameters which are found from the experimental data can be compared. The values of the different parameters should satisfy certain physically reasonable conditions which can serve as constraints to be used in addition to the question of whether the data can be fitted.

We will classify the one- and two-body operators here, but defer to Secs. 2 and 3 a detailed derivation of the mass formulas. We label the operators $T_{\dim SU(6)}{}^{\dim SU(3)}$. Since a single particle is in a **6** of $SU(6)$, the most general operator acting on a single particle must be in

$$\mathbf{6}\times\mathbf{6}^* = \mathbf{1}+\mathbf{35}.$$

Only $S=0$ operators can act on a single particle and thus the only operators available are $T_1{}^1$ and $T_{35}{}^8$, which, of course, essentially come from the central mass and the mass splitting between the nonstrange and strange quarks. [We didn't need this argument to find such a simple result, but gave it to indicate how this $SU(6)$ analysis will work.] For the two-body operators the most general state is in

$$\mathbf{6}\times\mathbf{6} = \mathbf{21}_{\rm sym}+\mathbf{15}_{\rm anti}.$$

The most general operators[18] acting on a symmetric space state are in

$$\mathbf{21}\times\mathbf{21}^* = \mathbf{1}+\mathbf{35}+\mathbf{405}.$$

The $S=0$ operators here are $T_1{}^1$, $T_{35}{}^8$, $T_{405}{}^1$, and $T_{405}{}^8$. The first two operators cannot be distinguished from the corresponding one-body operators in a system with a fixed number of particles and thus do not contribute anything new. The four $S=0$ operators acting on spatially symmetric two-body states lead precisely to the Gürsey-Radicati mass formula for the $(\mathbf{56},0^+)$, as we will show in Sec. 2. The same parameters determined by the $(\mathbf{56},0^+)$ are used in the $(\mathbf{70},1^-)$. The parameters, when properly normalized, are reduced two-body matrix elements. For the $(\mathbf{70},1^-)$, we must also consider the $S=0$ operators acting on the antisymmetric space state; these occur in

$$\mathbf{15}\times\mathbf{15}^* = \mathbf{1}+\mathbf{35}+\mathbf{189}.$$

The $S=0$ ones are $T_1{}^1$, $T_{35}{}^8$, $T_{189}{}^1$, and $T_{189}{}^8$. The four new terms introduced here have parameters which must be found from the experimental data. The spin-orbit interactions can act only on the antisymmetric two-body space wave function, since the relative angular momentum is antisymmetric under the permutations of the two particles. Thus only the $S=1$ operators in $\mathbf{15}\times\mathbf{15}^*$ can contribute, namely $T_{35}{}^1$, $T_{35}{}^8$, and $T_{189}{}^8$. Tensor forces can only act on the symmetric space state and thus occur only in $T_{405}{}^1$ and $T_{405}{}^8$.

The two-body dominance assumption leads to significant simplification. For example, from the standpoint of abstract $SU(6)$ the mass operator can occur in[16]

$$\mathbf{56}\times\mathbf{56}^* = \mathbf{1}+\mathbf{35}+\mathbf{405}+\mathbf{2695},$$

while from the two-body point of view the representation **2695** cannot occur. Similar simplifications occur for the states with $L>0$.

Section 2 gives a nonperturbative derivation of the Gürsey-Radicati mass for the $(\mathbf{56},0^+)$. Section 3 gives the analysis of the two-body operators for s- and p-wave particles; Sec. 4 gives the fit of the particles in the $(\mathbf{70},1^-)$, and Sec. 5 gives a summary and outlook for future work on this model.

2. DERIVATION OF THE GÜRSEY-RADICATI MASS FORMULA

We use a simple formalism with Bose operators to derive mass formulas in the symmetric-quark model. Let $a_\alpha{}^\dagger$ and a^α [$\alpha=Aa$, $A=1, 2, 3$ for $SU(3)$, $a=1, 2$ for $SU(2)_S$,[19] where S stands for ordinary spin] be a set of Bose creation and annihilation operators for s-wave quarks.[20] In terms of these operators, the generators of $SU(6)$ are

$$I_\alpha{}^\beta = a_\alpha{}^\dagger a^\beta - \tfrac{1}{6}\delta_\alpha{}^\beta N\,,\quad N = a_\gamma{}^\dagger a^\gamma.$$

Relevant subgroups of $SU(6)$ are given in Table I, where $N=N_n+N_\lambda$ is the quark-number operator, $N_n = a_{\bar A c}{}^\dagger a^{\bar A c}$ and $N_\lambda = a_{3c}{}^\dagger a^{3c}$ are the nonstrange and strange quark-number operators, respectively, and S_n and S_λ refer to the corresponding quark spins. The reduction chains which occur are $SU(6)\to SU(3)\times SU(2)_S$, and $SU(6)\to SU(4)\times SU(2)_{S_\lambda}$ with $SU(4)\to SU(2)_I\times SU(2)_{S_n}$.[21]

TABLE I. Generators of relevant subgroups of $SU(6)$.

Subgroup	Generators
$SU(3)$	$I_A{}^B = I_{Ac}{}^{Bc} = a_{Ac}{}^\dagger a^{Bc} - \frac{1}{3}\delta_A{}^B N$
$SU(2)_S$	$S_a{}^b = I_{Ca}{}^{Cb} = a_{Ca}{}^\dagger a^{Cb} - \frac{1}{2}\delta_a{}^b N$
$SU(2)_I$	$\mathcal{I}_{\bar A}{}^{\bar B} = I_{\bar A}{}^{\bar B} - \frac{1}{2}\delta_{\bar A}{}^{\bar B} I_{\bar E}{}^{\bar E} = a_{\bar A c}{}^\dagger a^{\bar B c} - \frac{1}{2}\delta_{\bar A}{}^{\bar B} Y$
$U(1)_Y$	$Y = I_{3c}{}^{3c} = I_{\bar E c}{}^{\bar E c} = I_{\bar E}{}^{\bar E} = \frac{1}{3}(N_n - 2N_\lambda)$
$SU(4)$	$I_{\bar A a}{}^{\bar B b} - \frac{1}{4}\delta_{\bar A}{}^{\bar B}\delta_a{}^b Y$
$SU(2)_{S_n}$	$I_{\bar A a}{}^{\bar B b} - \frac{1}{2}\delta_a{}^b Y$
$SU(2)_{S_\lambda}$	$I_{3a}{}^{3b} + \frac{1}{2}\delta_a{}^b Y$

[17] We keep all operators for (spin) $S=0$, and those in the lowest $SU(6)$ irreducible (**35**) for $S=1$.

[18] Conservation of parity prohibits operators connecting symmetric- and antisymmetric-space states.

[19] We will always use small Greek letters for $SU(6)$, capital Latin letters for $SU(3)$, and small Latin letters for $SU(2)_S$. Later we will use capital barred Latin letters for $SU(2)_I$. Repeated indices are to be summed.

[20] S-wave quarks suffice for the Gürsey-Radicati mass formula. When we discuss the $(\mathbf{70},1^-)$ in Sec. 3, we will extend the formalism to include p-wave quarks. Particles with higher l and higher principal quantum number can be treated in a similar way.

[21] This formalism is that given by M. A. B. Bég and V. Singh, Phys. Rev. Letters **13**, 418 (1964), translated into second quantized form using Bose operators.

Some other relevant objects are the Casimir operators for various $SU(n)$ groups. For example,

$$C_2^{(6)}=I_\alpha{}^\beta I_\beta{}^\alpha,\quad C_2^{(3)}=I_A{}^B I_B{}^A,$$
$$C_2^{(2)}(S)=S_a{}^b S_b{}^a=2S(S+1),$$
$$C_2^{(2)}(I)=\mathcal{G}_{\bar{A}}{}^{\bar{B}}\mathcal{G}_{\bar{B}}{}^{\bar{A}}=I_{\bar{A}}{}^{\bar{B}}I_{\bar{B}}{}^{\bar{A}}-\tfrac{1}{2}Y^2=2I(I+1),$$

where we added a distinguishing letter in those cases where the same $SU(n)$ group enters more than once.

Now we derive the Gürsey-Radicati formula. We ignore the one-body operators since they do not give anything which will not occur in the two-body operators. The two-body operators all have the form $a_\alpha{}^\dagger a_\beta{}^\dagger M_{\gamma\delta}{}^{\alpha\beta}a^\gamma a^\delta$, where the coefficients M must be determined to make the operator lie in given $SU(6)$ and $SU(3)$ irreducibles, satisfy octet dominance,[22] and have $S=0$. The method of construction of these operators is elementary and so we will just list them in Table II, together with their form in terms of quantum numbers.[23] The numerical matrix $y_A{}^B$ is diagonal and has diagonal elements $\frac{1}{3}$, $\frac{1}{3}$, and $-\frac{2}{3}$. The Gürsey-Radicati formula for the $(\mathbf{56},0^+)$ results follows:

$$M=M_0+M_1Y+M_2C_2^{(3)}+M_3[I(I+1)-\tfrac{1}{4}Y^2].\quad (1)$$

The Casimir operator $C_2^{(3)}$ can be eliminated in favor of $J(J+1)$ [$J=S$ in the $(\mathbf{56},0^+)$] to give a more familiar form of the Gürsey-Radicati mass formula.[24,25]

The Gürsey-Radicati mass formula predicts the following equalities:

$$N+\Xi=\tfrac{1}{2}(3\Lambda+\Sigma),\quad (2)$$

the Gell-Mann–Okubo formula for the octet;

$$\Omega-\Xi^*=\Xi^*-Y_1{}^*=Y_1{}^*-N^*=\Xi-\Sigma,\quad (3)$$

TABLE II. Forms of the two-body operators.

Operator	Second quantized form	Form in terms of quantum numbers
$T_1{}^1$	$a_\alpha{}^\dagger a_\beta{}^\dagger a^\alpha a^\beta$	$N(N-1)$
$T_{35}{}^8$	$y_B{}^A a_{Ac}{}^\dagger a_\mu{}^\dagger a^{Bc}a^\mu$	$(N-1)Y$
$T_{405}{}^1$	$a_{Aa}{}^\dagger a_{Bb}{}^\dagger a^{Ba}a^{Ab}-(5/7)a_\mu{}^\dagger a_\nu{}^\dagger a^\mu a^\nu$	$C_2^{(3)}-(8/21)N^2-(16/7)N$
$T_{405}{}^8$	$y_B{}^A[a_{Aa}{}^\dagger a_{Cb}{}^\dagger a^{Ca}a^{Bb}-\frac{7}{8}a_{Ac}{}^\dagger a_\mu{}^\dagger a^{Bc}a^\mu]$	$I(I+1)-\frac{1}{4}Y^2-\frac{1}{6}C_2^{(3)}-(5/24)(N+3)Y$

the equal spacing for the decuplet, and a relation between decuplet and octet. The average of the first three mass differences in Eq. (3) is 146 MeV and the last number is 124 MeV, so the best one can do in fitting the $(\mathbf{56},0^+)$ is to get within 11 MeV of all the masses.

3. ANALYSIS OF TWO-BODY FORCES FOR s- AND p-WAVE PARTICLES

With artificial single-particle space states, at first sight the following two-body parameters enter for s- and p-wave particles: s-s, $(s\text{-}p)_{\text{sym}}$, $(p\text{-}p)_{\text{sym},l=0}$, $(p\text{-}p)_{\text{sym},l=2}$, $(s\text{-}p)_{\text{anti}}$, and $(p\text{-}p)_{\text{anti}}$. Since with relative coordinates only two kinds of relative space states among s- and p-wave particles occur: $l=0$, which is symmetric in space, and $l=1$ which is antisymmetric in space, we must reduce the number of two-body parameters in some way. The simplest way to make this reduction is to identify all the $l=0$ cases and all the $l=1$ cases. The $l=2$ case does not contribute to the $(\mathbf{70},1^-)$ and $(\mathbf{20},1^+)$. This procedure has the virtues of being simple and of not introducing any more parameters than one would have in the more careful treatment with relative states. With this assumption the two-body parameters determined from the $(\mathbf{56},0^+)$ and the $(\mathbf{70},1^-)$ completely determine the $(\mathbf{20},1^+)$. The predictions about the $(\mathbf{20},1^+)$ which can then be made are a good check on this assumption; however, we expect the results for the $(\mathbf{70},1^-)$ to be more accurate than those for the $(\mathbf{20},1^+)$. Table III gives a summary of the parameters used and the resonances predicted.

To derive the mass formula for s- and p-wave particles, it is convenient to introduce Bose operators $c_{\alpha i}{}^\dagger$ and $c^{\alpha i}=(c_{\alpha i}{}^\dagger)^\dagger$, $i=0,1,2,3$, with

$$c_{\alpha i}{}^\dagger=a_\alpha{}^\dagger,\quad i=0$$
$$=b_{\alpha i}{}^\dagger,\quad i=1,2,3,$$

where the $a^\dagger$'s create s-wave particles as before and the

TABLE III. Summary of the parameters used and the resonances predicted.

Term	Fixed by	No. of parameters	No. of resonances
Symmetric, $S'=0$	$(\mathbf{56},0^+)$	4	8 in **56**, $L=0^+$
Antisymmetric, $S=0$	$(\mathbf{70},1^-)$	4	{30 in **70**, $L=1^-$
Spin-orbit	$(\mathbf{70},1^-)$	2	{11 in **20**, $L=1^+$

[22] Octet dominant operators either are a singlet under $SU(3)$ or lie in an octet and have $I=Y=0$.

[23] We give one sample of the manipulations leading from the second quantized form listed to the form in terms of the quantum numbers of the observed particles. Consider $T_{405}{}^8$. The irreducible **405** must be symmetric separately in upper and lower indices, and traceless:

$$(T_{405}{}^8)_{\alpha\beta}{}^{\gamma\delta}=a_\alpha{}^\dagger a_\beta{}^\dagger a^\gamma a^\delta-\tfrac{1}{8}(\delta_\alpha{}^\gamma\eta_\beta{}^\delta+\delta_\alpha{}^\delta\eta_\beta{}^\gamma+\delta_\beta{}^\delta\eta_\alpha{}^\gamma+\delta_\beta{}^\gamma\eta_\alpha{}^\delta)+(1/56)(\delta_\alpha{}^\gamma\delta_\beta{}^\delta+\delta_\alpha{}^\delta\delta_\beta{}^\gamma)\eta,$$

where $\eta=a_\mu{}^\dagger a_\nu{}^\dagger a^\mu a^\nu$ and $\eta_\alpha{}^\beta=a_\alpha{}^\dagger a_\mu{}^\dagger a^\beta a^\mu$. The $I=Y=0$ member of the $S=0$ octet is $Y_B{}^A(T_{405}{}^8)_{Aa,Cb}{}^{Ca,Bb}$, which simplifies to the form listed. The first term of the form listed is

$$y_B{}^A a_{Aa}{}^\dagger a_{Cb}{}^\dagger a^{Ca}a^{Bb}$$
$$=\tfrac{1}{2}y_B{}^A(a_{Aa}{}^\dagger a^{Ca}a_{Cb}{}^\dagger a^{Bb}-3\delta_b{}^a a_{Aa}{}^\dagger a^{Bb}+a_{Cb}{}^\dagger a^{Bb}a_{Aa}{}^\dagger a^{Bb})$$
$$=\tfrac{1}{2}y_B{}^A[I_A{}^C+\tfrac{1}{3}\delta_A{}^C N,\ I_C{}^B+\tfrac{1}{3}\delta_C{}^B N]_+-\tfrac{3}{2}Y.$$

The result listed is obtained by using the identity

$$I_A{}^C I_C{}^A=I_{\bar{A}}{}^{\bar{E}}I_{\bar{E}}{}^{\bar{A}}+[I_A{}^3,I_3{}^{\bar{A}}]_++(I_3{}^3)^2$$

and adding the second term.

[24] Using the Bose operators, it is straightforward to show that $C_2^{(3)}=C_2^{(2)}(S)+\frac{1}{6}N^2+N$ for symmetric representations of $SU(6)$. For $N=3$, this reduces to $C_2^{(3)}=2S(S+1)+\frac{9}{2}$, valid for the **56**. This last result was first given by Bég and Singh, Ref. 21.

[25] P. Federman, H. R. Rubenstein, and I. Talmi [Phys. Letters **22**, 208 (1966)] studied the **56** from the point of view of two-body dominance, but did not use the $SU(6)$ symmetry systematically.

$b^\dagger$'s create p-wave particles with $i=1, 2, 3$ corresponding to magnetic quantum number $-1, 0, 1$, respectively. The generators are similar to those given above, with the c's replacing the a's. For example, the generators of $SU(6)$ are

$$I_\alpha{}^\beta = c_{\alpha i}{}^\dagger c^{\beta i} - \tfrac{1}{6}\delta_\alpha{}^\beta N, \quad N = c_{\mu i}{}^\dagger c^{\mu i}.$$

The one-body operator is

$$M_n N_n + M_\lambda N_\lambda = (\tfrac{2}{3}M_n + \tfrac{1}{3}M_\lambda)N + (M_n - M_\lambda)Y,$$

where in a naive model M_n and M_λ would be the nonstrange and strange quark masses. The $S=0$ terms can again be manipulated into the form of quantum numbers; however now quantum numbers for both the $SU(3)$ and $SU(4)$ chains occur, and the observed resonances are, in general, not eigenstates in either chain. In Table IV we list the two-body operators in terms of quantum numbers. The second quantized forms from which they were derived can be written down easily in analogy with those given above for the $(\mathbf{56},0^+)$. The new superscript on the left indicates the $SU(6)$ two-body state on which they act. The $S=0$ mass formula which results from the terms given in Table IV is

$$M = \sum M_i \mathfrak{N}_i T_i,$$

where the normalization factors

$$\mathfrak{N}_i = (T_{i,\max} - T_{i,\min} + 1)^{-1}$$

are chosen to make all the terms contribute in comparable ways in the two-body system; the M's are then reduced two-body matrix elements. Note that for $T=I_3$, the third component of the isospin, the factor becomes the standard $(2I+1)^{-1}$. These normalization factors are, in the order given in Table IV: $\frac{1}{2}$, $\frac{1}{3}$, $\frac{1}{5}$, $9/28$, $\frac{1}{2}$, $\frac{1}{3}$, $\frac{1}{5}$, $\frac{2}{5}$. Notice that for a symmetric representation the terms acting on the **21** reduce to those considered in Sec. 2, and the terms acting on the **15** vanish. For later numerical work it is convenient to write the $S=0$ mass formula in the form[26,27]

$$\begin{aligned} M = {} & N_0 + N_1 Y + N_2[C_2^{(3)} + 2S(S+1)] \\ & + N_3[I(I+1) - \tfrac{1}{4}Y^2 + S_n(S_n+1) - S_\lambda(S_\lambda+1)] \\ & + N_4(C_2^{(6)} - 45/2) + N_5[Y - \tfrac{3}{4} + I(I+1) - \tfrac{1}{4}Y^2 \\ & - S_n(S_n+1) + S_\lambda(S_\lambda+1)] \\ & + N_6[C_2^{(3)} - 2S(S+1) - \tfrac{9}{2}] \\ & + N_7[C_2^{(4)} - 2S_\lambda(S_\lambda+1) - \tfrac{1}{4}Y^2 - 8Y - 15/2], \end{aligned} \quad (4)$$

where N_4 through N_7 inclusive vanish for the $(\mathbf{56},0^+)$.

[26] Except for the terms with $C_2^{(6)}$, these terms occur in Bég and Singh, Ref. 21. Our two-body dominance assumption provides a rationale for the particular choice of $SU(6)$ irreducible operators made by Bég and Singh. We also relate different $(SU(6),L)$ supermultiplets. Bég and Singh [Phys. Rev. Letters **13**, 509 (1964)] discuss the **70** in abstract $SU(6)$ (without orbital excitation).

[27] A. W. Hendry [Nuovo Cimento 48, 780 (1967)] studied the **56**, **70**, and **20** in a way similar to Federman *et al.*, Ref. 25.

Table IV. The two-body operators in terms of quantum numbers.

Operator	Form in terms of quantum numbers
${}^{21}T_1{}^1$	$\frac{1}{2}[C_2^{(6)} + (7/6)N^2 - 7N]$
${}^{21}T_{35}{}^8$	$\frac{1}{4}[C_2^{(4)} - 2S_\lambda(S_\lambda+1) - \frac{1}{4}Y^2 - \frac{1}{3}C_2^{(6)}] + 2(\frac{1}{3}N - 1)Y$
${}^{21}T_{405}{}^1$	$\frac{1}{2}[C_2^{(3)} + 2S(S+1) - (5/7)C_2^{(6)}]$
${}^{21}T_{405}{}^8$	$\frac{1}{2}[I(I+1) - \frac{1}{4}Y^2 + S_n(S_n+1) - S_\lambda(S_\lambda+1)] - \frac{1}{12}[C_2^{(3)} + 2S(S+1)] - \frac{7}{32}[C_2^{(4)} - 2S_\lambda(S_\lambda+1) - \frac{1}{4}Y^2 - \frac{1}{3}C_2^{(6)}] + 21/4(N/3-1)Y$
${}^{15}T_1{}^1$	$\frac{1}{2}(-C_2^{(6)} + \frac{5}{6}N^2 + 5N)$
${}^{15}T_{35}{}^8$	$-\frac{1}{4}[C_2^{(4)} - 2S_\lambda(S_\lambda+1) - \frac{1}{4}Y^2 - \frac{1}{3}C_2^{(6)}] + (\frac{1}{3}N + 1)Y$
${}^{15}T_{189}{}^1$	$\frac{1}{2}[C_2^{(3)} - 2S(S+1)] - \frac{1}{10}C_2^{(6)}$
${}^{15}T_{189}{}^8$	$\frac{1}{2}[I(I+1) - \frac{1}{4}Y^2 - S_n(S_n+1) + S_\lambda(S_\lambda+1)] - \frac{1}{12}[C_2^{(3)} - 2S(S+1)] + \frac{1}{16}[C_2^{(4)} - 2S_\lambda(S_\lambda+1) - \frac{1}{4}Y^2 - \frac{1}{3}C_2^{(6)}]$

The N_i are related to the one- and two-body parameters introduced above by

$$\begin{aligned} \tfrac{2}{3}M_n + \tfrac{1}{3}M_\lambda - (21/8)({}^{21}M_1{}^1) + (45/8)({}^{15}M_1{}^1) \\ = N_0 - (45/2)N_4 - \tfrac{3}{4}N_5 - \tfrac{9}{2}N_6 - (15/2)N_7, \end{aligned}$$

$$M_n - M_\lambda + \tfrac{2}{3}({}^{15}M_{35}{}^8) = N_1 + N_5 - 8N_7,$$

$$\begin{aligned} ({}^{21}M_1{}^1) - ({}^{15}M_1{}^1) = 4N_4 + (20/7)N_2 + (10/21)N_3 \\ + (2/15)N_5 + \tfrac{4}{5}N_6 + \tfrac{4}{3}N_7, \end{aligned}$$

$$({}^{21}M_{35}{}^8) - ({}^{15}M_{35}{}^8) = 12N_7 + (21/4)N_3 - \tfrac{3}{2}N_5,$$

$$({}^{21}M_{405}{}^1) = 10N_2 + (5/3)N_3,$$

$$({}^{21}M_{405}{}^8) = (56/9)N_3,$$

$$({}^{15}M_{189}{}^1) = 10N_6 + (5/3)N_5,$$

$$({}^{15}M_{189}{}^8) = 5N_5.$$

Note that only the **405** and **189** two-body parameters can be separated individually from the one-body parameters.

The $SU(6)$-symmetric terms $\alpha + \beta C_2^{(6)}$, above, lead to equal spacing among the $(\mathbf{56},0^+)$, the $(\mathbf{70},1^-)$ and the $(\mathbf{20},1^+)$ in increasing order.

The spin-orbit two-body operators cannot conveniently be put in the form of quantum numbers in the $SU(3)$ or $SU(4)$ chains because the spin-orbit operator breaks the $SU(6)$ L-S coupling model and has off-diagonal terms.[28] These terms can be evaluated either directly from their second quantized expressions, which we give below, or using standard $3j$ and $6j$ techniques. This is the only place in which the states in the $(\mathbf{70},1^-)$ are needed; since these states are easily written down, we will not tabulate them here. We computed these independently by both methods as a check on the calculation. We kept the $S=1$ operators in the **35**, since this is the smallest irreducible of $SU(6)$ from which spin-orbit operators can be constructed. Recalling that spin-orbit operators act only on the **15**, and using the

[28] We keep the off-diagonal matrix elements of the spin-orbit operators in the $(\mathbf{70},1^-)$; however, we drop the matrix elements of these operators to other supermultiplets (configuration mixing).

standard $l=1$ matrices,[29] we have[30]

$$T_{L}\cdot s^{1}=(b_{Aa i}{}^{\dagger}a_{\gamma}{}^{\dagger}-b_{\gamma i}{}^{\dagger}a_{Aa}{}^{\dagger})(b^{Abj}a^{\gamma}-b^{\gamma j}a^{Ab})\sigma_{b}{}^{a}\cdot l_{j}{}^{i}, \quad (5)$$

and

$$T_{L}\cdot s^{8}=y_{B}{}^{A}(b_{Aai}{}^{\dagger}a_{\gamma}{}^{\dagger}-b_{\gamma i}{}^{\dagger}a_{Aa}{}^{\dagger})\times(b^{Bbj}a^{\gamma}-b^{\gamma j}a^{Bb})\sigma_{b}{}^{a}\cdot l_{j}{}^{i}. \quad (6)$$

We write the total spin-orbit operator

$$N_{8}T_{L}\cdot s^{1}+N_{9}T_{L}\cdot s^{8}, \quad (7)$$

and give these operators in the $SU(3)$ chain, in the Appendix.

The $S=2$ (tensor forces) act only on the symmetric space states and occur only in the **405** of $SU(6)$. We drop these terms here because there is not enough data at present to evaluate them. It is worth pointing out that only these terms resolve the degeneracy between the $J=\frac{1}{2}$ and $\frac{3}{2}$, Δ and Ω decuplet resonances; conversely, the splittings of those resonances will determine the octet-dominant tensor forces.

4. ANALYSIS OF DATA FOR RESONANCES IN THE (70,1⁻)

The first step in the analysis is to determine where negative-parity baryon resonances with given J, I, and Y quantum numbers can fit in the $(\mathbf{70},1^{-})$. The possible placement of resonances is shown in Table V, where the expression under placement is $(\dim SU(3), \dim SU(2)_S)$. Since there are not yet enough uniquely placed resonances known experimentally, we used the following experimental indications to place some of the non-unique resonances: the quartet octet lies higher than the doublet octet, and the doublet octet lies higher than the singlet.[31] We used the table of Rosenfeld *et al.*[32] for the experimental data. Their table lists nine particles which they consider well determined, whose quantum numbers are such that they could fit in the $(\mathbf{70},1^{-})$; however, two of these might be S-wave threshold effects. The remaining seven particles are listed in Table VI, assigned as described above. It should be emphasized that later on we will allow mixing among all the resonances that can mix.

TABLE V. Possible placement of resonances in $(\mathbf{70},1^{-})$.

Resonance	J^P	Placement	Unique?
N, Λ, Σ, Ξ	$\frac{5}{2}^-$	(8,4)	Yes
Δ, Ω	$\frac{1}{2}^-$, $\frac{3}{2}^-$	(10,2)	Yes
N	$\frac{1}{2}^-$, $\frac{3}{2}^-$	(8,2), (8,4)	No
Λ	$\frac{1}{2}^-$, $\frac{3}{2}^-$	(1,2), (8,2), (8,4)	No
Σ, Ξ	$\frac{1}{2}^-$, $\frac{3}{2}^-$	(8,2), (10,2), (8,4)	No

TABLE VI. Well-determined resonances in the $(\mathbf{70},1^{-})$.

Resonance	J^P	Placement	Unique?
Λ(1405)	$\frac{1}{2}^-$	(1,2)	No
Λ(1519)	$\frac{3}{2}^-$	(1,2)	No
N(1525)	$\frac{3}{2}^-$	(8,2)	No
N(1570)	$\frac{1}{2}^-$	(8,2)	No
N(1670)	$\frac{5}{2}^-$	(8,4)	Yes
Δ(1670)	$\frac{1}{2}^-$	(10,2)	Yes
Σ(1768)	$\frac{5}{2}^-$	(8,4)	Yes

For the symmetric space parameters we kept the parameters that are determined by the **56**. We determined the antisymmetric space parameters as follows. First we neglected spin-orbit forces and estimated the $S=0$ antisymmetric space parameters from the central masses of the L-S multiplets. We used $N(1525)$, $\Delta(1670)$, $\Sigma(1768)$, and estimated the fourth parameter by assuming $M_{189}{}^{8}/M_{189}{}^{1}\simeq M_{405}{}^{8}/M_{405}{}^{1}$. Since there are more resonances than parameters, we predicted the remaining particles from the parameters we had chosen, neglecting, of course, spin-orbit terms. These predictions, which took account of the $SU(3)$ mixing between the Λ's in (1,2) and (8,2), agreed rather well with the experimental data. We then estimated the two spin-orbit parameters from the large splitting between the singlet Λ's with spin $\frac{1}{2}$ and $\frac{3}{2}$, keeping the $SU(3)$ mixings which had been determined in the earlier step for these Λ's. We used these spin-orbit parameters to predict the spin-orbit splittings in the quartet octet; the results were encouraging but could be improved by adjusting the spin-orbit parameters, which we did. At this step of the calculation, we programmed all of the particles in the $(\mathbf{70},1^{-})$ including diagonalizations of all the mixings [both $SU(3)$ and spin mixings]. We arranged the calculation so that inserting the six parameters to be varied gave the masses of all the isospin multiplets in the $(\mathbf{70},1^{-})$. In order to find the best set of parameters, we introduced a figure of merit equal to the sum of the squares of the differences between the calculated and experimental masses for the seven well-established particles. We then searched, using the computer, for the values of those six parameters which minimized the figure of merit. During this search, we found no evidence for the existence of local minima; however, we did not attempt to exclude this possibility in a systematic way. The final figure of merit was 286, which means that in the least-squared sense the theoretical values are on the average 6.4 MeV from the experimental values. In fact, the worst fit particle, the Λ(1520), was 11 MeV away from the theoretical value and contributed almost half of the total figure of merit. By comparison with the accuracy of the Gürsey-Radicati formula for the $(\mathbf{56},0^{+})$ this fit is good. We want to emphasize again that *a priori* there is no

[29] L. I. Schiff, *Quantum Mechanics* (McGraw-Hill Publishing Company, Inc., New York, 1955), p. 146.

[30] These expressions are valid for the $(\mathbf{70},1^{-})$ where the spin-orbit terms act on an $(s\text{-}p)_{\text{anti}}$ space state.

[31] Strictly speaking, this holds only for the N and Λ for which the two octets mix (together with the singlet for the Λ) but the decuplet does not enter.

[32] A. H. Rosenfeld, *et al.*, Rev. Mod. Phys. **39**, 1 (1967).

TABLE VII. Calculation versus experiment for **(70,1⁻)**. Asterisk stands for well-established input resonances; question mark stands for resonances less well established; superscript M stands for resonances mixed by more than 20%. The left columns are calculated masses. The right columns are experimental masses.

	^4P $J=\frac{5}{2}$	$\frac{3}{2}$	$\frac{1}{2}$		^2P $\frac{3}{2}$	$\frac{1}{2}$
				Ω	2031	2031
				Ξ	1851^M, 1815?	1850^M
Ξ	1853	1705^M	1701^M	Σ	1759^M	1692^M
Z	1763, 1768*	1572^M	1612^M	Δ	1676	1676, 1670*
Λ	1763	1808	1815			
N	1672, 1670*	1744	1784	Ξ	1954^M, 1933?	1799^M
				Σ	1798^M	1827^M
				Λ	1635, 1682?	1705^M, 1670?
				N	1520, 1525*	1563, 1570*
				Λ	1531, 1519*	1402^M, 1405*

guarantee that one could fit the intricate spin-orbit splitting of the particles with only two spin-orbit parameters. Table VII lists the calculated masses along with the well-established experimental particles and some other particles that are not as well-established.[33]

TABLE VIII. Mixing amplitudes for resonances in the **(70,1⁻)**. The N's are essentially unmixed.

Particle (mass) (J)	Amplitudes: (S,SU(3))		
	($\frac{1}{2}$,1)	($\frac{1}{2}$,8)	($\frac{3}{2}$,8)
Λ(1402) ($\frac{1}{2}$)	0.76	−0.65	0.02
Λ(1705) ($\frac{1}{2}$)	−0.62	−0.71	0.35
Λ(1815) ($\frac{1}{2}$)	0.21	0.27	0.94
Λ(1531) ($\frac{3}{2}$)	−0.99	0.14	0.08
Λ(1635) ($\frac{3}{2}$)	−0.10	−0.94	0.33
Λ(1808) ($\frac{3}{2}$)	0.12	0.32	0.94
	($\frac{1}{2}$,10)	($\frac{1}{2}$,8)	($\frac{3}{2}$,8)
Σ(1612) ($\frac{1}{2}$)	−0.63	0.03	0.78
Σ(1692) ($\frac{1}{2}$)	0.70	0.47	0.54
Σ(1827) ($\frac{1}{2}$)	0.35	−0.88	0.32
Σ(1572) ($\frac{3}{2}$)	−0.67	0.37	0.64
Σ(1759) ($\frac{3}{2}$)	0.72	0.54	0.44
Σ(1798) ($\frac{3}{2}$)	0.19	−0.75	0.63
Ξ(1701) ($\frac{1}{2}$)	−0.57	−0.31	0.76
Ξ(1799) ($\frac{1}{2}$)	0.35	0.74	0.57
Ξ(1850) ($\frac{1}{2}$)	0.74	−0.60	0.31
Ξ(1705) ($\frac{3}{2}$)	−0.70	0.06	0.71
Ξ(1851) ($\frac{3}{2}$)	0.60	−0.50	0.63
Ξ(1954) ($\frac{3}{2}$)	0.39	0.87	0.31

[33] V. S. Bhasin, D. L. Katyal, and A. N. Mitra [Phys. Rev. **161**, 1546 (1967)] also studied the **(70,1⁻)** in the symmetric quark model. Although they used the ideas of two-body and octet dominance, they did not try to find the most general mass operator with these properties, and did not make an $SU(6)$ analysis of their operators. They did not find a quantitative fit to the well-established resonances. They assigned N(1570) to (8,4) citing Dalitz's analysis, Ref. 4, and the decay analysis of Mitra and Ross, Ref. 34, while we assign it to (8,2) on the basis of its mass. The (8,4) assignment for N(1570) depresses the predicted masses of the other resonances.

Note that none of the undiscovered resonances have mass less than 1550 MeV. We want to call particular attention to the following predictions: a $\Sigma(\frac{3}{2}^-)$ at 1572 MeV, an accidentally degenerate $\Lambda(\frac{5}{2}^-)$ at the same mass as the known $\Sigma(\frac{5}{2}^-)$ at 1768 MeV, and $\Xi(\frac{1}{2}^-)$ and $\Xi(\frac{3}{2}^-)$ at 1701 and 1705 MeV, respectively. This calculation at the same time produced the mixing amplitudes which are tabulated in the $SU(3)$ basis[34,35] in Table VIII. We remind the reader that the mixing probabilities are the squares of these numbers. The mixing amplitudes can be compared with experiment using branching ratios for various decays.

Our analysis resulted in values of the mass parameters which we introduced. Those for the antisymmetric space terms are generally comparable in magnitude but smaller than the similar parameters for the symmetric space terms. The results are[36]

$$({}^{21}M_1{}^1)-({}^{15}M_1{}^1)=-291.5\text{ MeV},$$

$$({}^{21}M_{35}{}^8)-({}^{15}M_{35}{}^8)=-22.9\text{ MeV},$$

$$M_{405}{}^1=199\text{ MeV},\quad M_{405}{}^8=121\text{ MeV},$$

$$M_{189}{}^1=36.8\text{ MeV},\quad M_{189}{}^8=-103\text{ MeV}.$$

The spin-orbit parameters, which for convenience are normalized using the diagonal terms of their contributions to the observed resonance states in the $SU(3)$ basis, are

$$M_{L\cdot S}{}^1=35.4\text{ MeV},\quad M_{L\cdot S}{}^8=-189\text{ MeV}.$$

It is interesting to note that the octet term has opposite sign and is about six times larger in magnitude than the singlet term.

We emphasize that the strong mixture among $J^P=\frac{1}{2}^-$ and $\frac{3}{2}^-$ resonances in the $(SU(3), SU(2)_S)$ basis implies that the Gell-Mann–Okubo mass formula will not be valid, and that it makes no sense to try to group these particles into octets or decuplets.

[34] A. N. Mitra and M. Ross [Phys. Rev. **158**, 1630 (1967)] studied decays of the **(70,1⁻)** in the symmetric quark model. We will make a few comments about such decays, even though we have not made a systematic analysis of them. Mitra and Ross find a kinematic effect enhancing S-wave decay into high-mass mesons (κ and η over π) which provides a mechanism to break $SU(3)$ in decays; however, they do not introduce an explicit octet-dominant $SU(3)$-violating term in the decay interaction. We suggest that such $SU(3)$-violating terms be allowed, and that their magnitude, relative to the $SU(3)$-conserving ones be determined when the experimental data is sufficient. The spin-orbit terms which we found give one example where the (octet-dominant) $SU(3)$-violating terms are larger than the $SU(3)$-conserving ones. (See footnote 35.)

[35] G. B. Yodh [Phys. Rev. Letters **18**, 810 (1967)] analyzed the decays of Λ(1519) assuming strict $SU(3)$ symmetry and found a conflict between the singlet assignment and the experimental branching ratios. Our analysis suggests that an $SU(3)$-violating term in the decay interaction is necessary to resolve this conflict [Yodh's interpretation (b)].

[36] The N_i introduced in Eqs. (4) and (7) are, in order (and in MeV): 997.5, −168.1, 16.69, 19.45, −79.95, −20.51, 7.1, −12.98, 2.08, and −26.99.

5. SUMMARY AND OUTLOOK

We studied the symmetric quark model of baryons (in which the baryons are composites of three quarks with wave functions symmetric in the visible quantum numbers) with orbital excitation of quarks using approximate phenomenological $SU(6)$ symmetry with spin-orbit coupling to resolve the degeneracy among resonances with the same L and S but different J. To reduce the number of possible mass formulas, we assumed dominance of two-body contributions to the mass operator in addition to the octet-dominance which is usually a part of an approximately $SU(3)$-symmetric theory. We showed that octet- and two-body dominance yield the Gürsey-Radicati mass formula for the $(\mathbf{56},0^+)$, thereby giving a derivation of this formula free of perturbation theory. The same principles, together with dominance of spin-orbit operators lying in the lowest possible $SU(6)$ irreducible, lead to a ten-parameter unified mass formula for the 49 isospin multiplets lying in the $(\mathbf{56},0^+)$, $(\mathbf{70},1^-)$, and $(\mathbf{20},1^+)$. The large $SU(6)$ symmetric terms in this formula obey an equal spacing rule which places the $(\mathbf{20},1^+)$ above 2 BeV, so that the $N(1400)$ resonance cannot be placed in the $(\mathbf{20},1^+)$. We found a quantitative fit for the seven well-established negative-parity baryon resonances which can be placed in the $(\mathbf{70},1^-)$, and predicted the remaining resonances in this supermultiplet. For this fit the octet spin-orbit term was (in magnitude) about six times as large as the singlet one (and opposite in sign).

A table of higher supermultiplets in the symmetric quark model with orbital excitation was given earlier.[12] Here we emphasize that the $(\mathbf{56},1^-)$ is spurious and does not occur in the quark model, so that the $(\mathbf{56},2^+)$ is the next **56** above the $(\mathbf{56},0^+)$, just as in the Regge theory. Several authors have pointed out that the $\Sigma(2035)$ and $\Delta(1920)$ fit in the ${}^4D_{7/2}$, and $\Sigma(1910)$, $\Lambda(1820)$ and $N(1688)$ fit in the ${}^2D_{5/2}$ of this supermultiplet. Only tensor forces split the degenerate J multiplets here. It seems premature to make a detailed study of the mass formula for this case.

We point out an interesting regularity among nucleon resonances which deserves further study. It is well-known that there are two series of resonances of opposite parity: $\Delta((\frac{3}{2}+2n)^+)$, $n=0, 1, \cdots 4$, and $N((\frac{3}{2}+2n)^-)$, $n=0, 1, \cdots 3$, where the number in parentheses is J. In the symmetric quark-quark model these opposite parity series can be united into one family. Place the Δ's in the series (dim $SU(6)$, L^P)$=(\mathbf{56},(2n)^+)$, $n=0, 1, \cdots 4$, and the N's in the series $(\mathbf{70}, (2n+1)^-)$.[37] Then a plot of M^2 versus L shows that both series lie on a single straight line with equation

$$M^2=1.46+1.09L \text{ in } (\text{BeV})^2.$$

This regularity suggests that something like the exchange degeneracy which was found for mesons[38] may also occur for baryons.

When more supermultiplets have been analyzed, a systematic analysis of the reduced two-body parameters classified by $SU(6)$ may be useful in revealing regularities in the interactions.

APPENDIX: SPIN-ORBIT OPERATORS

We list in Tables IX and X the (real symmetric) matrices of the spin-orbit operators $T_{L\cdot S^1}$ and $T_{L\cdot S^8}$ in the $SU(3)$ chain. The properly normalized spin-orbit operator is

$$(1/17)M_{L\cdot S^1}T_{L\cdot S^1}+(1/7)M_{L\cdot S^8}T_{L\cdot S^8},$$

with

$$M_{L\cdot S^1}=17N_8, \quad M_{L\cdot S^8}=7N_9.$$

TABLE IX. Matrix elements of $T_{L\cdot S^1}$.

	$J=\frac{5}{2}$	$\frac{3}{2}$	$\frac{1}{2}$		$J=\frac{3}{2}$	$\frac{1}{2}$
(8,4)	6	-4	-10			
(**10**,2)		0	0	$\langle 8,4\vert T_{L\cdot S^1}\vert 8,2\rangle$	$-\sqrt{10}$	-2
(8,2)		2	-4			
(**1**,2)		4	-8			

TABLE X. Matrix elements of $T_{L\cdot S^8}$. $T_{L\cdot S^8}=0$ on the Δ and Ω states. $T_{L\cdot S^8}=\frac{1}{3}T_{L\cdot S^1}$ on the nucleon states.

Basis states		$J=\frac{3}{2}$			$J=\frac{1}{2}$	
Λ(8,4)	$-\frac{2}{3}$	$-\frac{2}{3}\sqrt{10}$	$-\frac{1}{3}\sqrt{10}$	$-5/3$	$-\frac{4}{3}$	$-\frac{2}{3}$
Λ(8,2)		--	5/3		$\frac{4}{3}$	$-10/3$
Λ(**1**,2)			0			0
Σ(8,4)	$\frac{2}{3}$	$\frac{2}{3}\sqrt{10}$	$-\sqrt{10}$	5/3	$\frac{4}{3}$	-2
Σ(8,2)		$\frac{2}{3}$	-1		$-\frac{4}{3}$	2
Σ(**10**,2)			0			0
Ξ(8,4)	$\frac{3}{2}$	$-\frac{1}{3}\sqrt{10}$	$-\sqrt{10}$	5/3	$-\frac{2}{3}$	-2
Ξ(8,2)		$-\frac{4}{3}$	-1		8/3	2
Ξ(**10**,2)			0			0

[37] We ignore complications due to admixture of spurious states here.

[38] R. C. Arnold, Phys. Rev. Letters **14**, 657 (1965).

Reprinted from: RECENT DEVELOPMENTS IN HIGH-ENERGY PHYSICS
Edited by Behram Kursunoglu,
Arnold Perlmutter, and Linda F. Scott
(Plenum Publishing Corporation, 1980)

QUARKS IN LIGHT BARYONS*

Nathan Isgur

University of Toronto, Toronto, Ontario, Canada and

Gabriel Karl

University of Guelph, Guelph, Ontario, Canada
(presented by Gabriel Karl)

ABSTRACT: Recent work on quark models for baryons based on QCD is reviewed briefly.

I. INTRODUCTION, ASSUMPTIONS AND MODEL FOR BARYONS

It is now optimistically believed that all low energy particle physics phenomena, say below 5 GeV, are basically understood in the same rough sense in which Atomic Physics was understood in 1930. In the particular case of the hadron sector, Quantum Chromodynamics (QCD) is believed to be the basic theory. It is not certain as yet whether this belief is warranted, since exact results based rigorously on QCD are in short supply. In the absence of such exact results one has to make some guesses and hope that these guesses will be substantiated later from QCD. The work on baryons discussed here is mainly in this fashion of guess work mainly about the nature of interquark interactions.

*Work supported by N.S.E.R.C. Ottawa

There has been a great deal of recent work on hadrons primarily under the stimulation of the experimental discovery of J/ψ, Υ and related particles. The first theoretical wave of attack was directed at explaining the charmonium and other heavies. The principles developed for charmonia have been applied also to lighter hadrons, mesons and baryons, even though the use of nonrelativistic models is much harder to justify in these cases. Therefore we are discussing "post-charmonium" baryon models.

Relative to the pioneering work of Greenberg, Dalitz and their coworkers on baryons, there are new physical assumptions which restrict the range of possible interactions between quarks. These assumptions refer to properties of the long range 'confinement' and short range quark quark potential. The long range interaction is assumed to be spin independent and more importantly flavor independent. This assumption is similar to the universality of the Coulomb interaction in positronium and muonium. In the hadronic sector, the more familiar case is the assumption of a universal potential in J/ψ (charmonium) and in Υ (bottomonium), which is based in turn on the assumption that all quarks are color triplets. We similarly assume that a single (Lorentz scalar) potential function between quarks describes all baryons, irrespective of the number of strange quarks they contain. We take for this function a perturbed harmonic potential, with the perturbation treated in first order only. This choice is made for computational convenience.

The short range interaction is assumed appropriate to massless vector exchange, very similar, except for overall normalization to single photon exchange. The spin dependent part of this interaction, the so called 'hyperfine interaction' between quarks, first advocated by de Rujula et al. plays an important role also in light hadrons. The choice of (Lorentz) scalar confining potential has consequences also for the spin dependent interactions -- it leads to a large cancellation of spin-orbit forces -- as emphasized especially by Schnitzer.

The Hamiltonian for baryons reviewed here is

$$H = \sum_{i=1}^{3} m_i + \sum \frac{\vec{p}_i^{\,2}}{2m_i} + \sum_{i<j} V_{ij}^{conf}(r_{ij}) + \sum_{i<j} V_{hyp}^{ij} ,$$

$$V_{ij}^{conf}(r_{ij}) = \frac{1}{2} k\, r_{ij}^2 + \bar{V}(r_{ij}) ,$$

$$V_{hyp}^{ij} = \frac{2\alpha_s}{m_i m_j} \left[\frac{8\pi}{3} \vec{S}_i \cdot \vec{S}_j \, \delta^3(\vec{r}_{ij}) + \frac{1}{r_{ij}^3} (3\, \vec{S}_i \cdot \hat{r}_{ij} \vec{S}_j \cdot \hat{r}_{ij} - \vec{S}_i \cdot \vec{S}_j) \right] .$$

The idea of universality, or flavor independence consists in the confinement potential V_{ij}^{conf} being the same for any choice of quark masses.

II. DISCUSSION OF SOME SPECTROSCOPIC RESULTS

Given the choice of Hamiltonian it is a straightforward matter to find its eigenstates and corresponding eigenvalues in the various sectors - nonstrange, strangeness minus one, minus two etc. These results have been discussed in the literature and are too lengthy to describe in great detail here. Instead we only emphasize some general features of these results. These are:

i) Mode splitting in excited hyperons. This phenomenon corresponds to a shift in frequency dependent on the type of mode: if the strange quark is excited in a mode the frequency is lower than in a similar mode with the strange quark at rest. The simplest example is the splitting between $\Lambda 5/2^-$ and $\Sigma 5/2^-$, which arises as a consequence of this splitting. In the $\Lambda 5/2^-$ it is the nonstrange quarks which are excited, while in $\Sigma 5/2^-$ it is the strange quark which is excited, thus accounting for most of the splitting between these two states. This phenomena is somewhat similar to isotope shifts in molecules which arise because the forces between nuclei are the same in two different isotopes and only the masses change between them.

The selection of modes which depend on which quark is moving, strange or nonstrange is a particularly simple way of SU_3 breaking, which is strongly supported by the data. In the case of the P-wave 70 plet the basis chosen is of the type $|1\rangle \pm |8\rangle$ or $|10\rangle \pm |8\rangle$; in other words the axes are at 45^0 relative to the SU_3 axes. To take one simple example, the well known analysis of Hey, Litchfield and Cashmore, gave as the empirical composition of the $\Lambda 1/2^-(1405)$:

$$\text{(A)} \qquad |\Lambda(1405)\rangle = .85|^2 1\rangle + .46|^2 8\rangle + .25|^4 8\rangle \quad .$$

If we reexpress this state in terms of the rotated basis

$$|^2\rho\rangle = \frac{1}{\sqrt{2}} \left(|^2 1\rangle - |^2 8\rangle\right) \quad ,$$

$$|^2\lambda\rangle = \frac{1}{\sqrt{2}} \left(|^2 1\rangle + |^2 8\rangle\right) \quad ,$$

$$|^4\rho\rangle = |^4 8\rangle \quad ,$$

we find

$$\text{(B)} \qquad |\Lambda(1405)\rangle = .93|^2\lambda\rangle + .27|^2\rho\rangle + .25|^4\rho\rangle$$

which shows that this state is almost pure $^2\lambda$ as one might also expect from its low mass. The comparison of (A) with (B) shows how much more convenient it is to use the ρ, λ basis to express experimental findings when compared to the SU_3 basis (A) which is far from the physical states. Similar observations can be made in the case of the positive parity excited baryons as well as the other strangeness sector $S = -2$.

ii) Spin dependent doublet mixings. A glance at the state (A) (or its other expression (B)) shows that there is a relatively small but nonzero mixing between spin doublet and spin quartet states. This is also the case with states in the nonstrange sector where we do not have to worry about mode splitting due to a strange quark. It is relatively easy to show that the hyperfine interaction accounts quantitatively for the observed doublet-quartet mixing in excited baryons. This mixing accounts for the decays of P-wave baryons which constituted a problem for quark models until the quark hyperfine interactions were taken into account.

III. SU_6 BREAKING

Other consequences of quark hyperfine interactions are related to the breaking of SU_6. This involves e.g. mixing into the ground state baryons (which have a symmetric orbital wavefunction) of excited baryons with a 'mixed' orbital wavefunction. The consequences are quite remarkable.

For example, there is a theorem of Fishbane et al. which says that the neutron charge radius must vanish as long as the neutron is a 56-plet. Hyperfine interactions, as shown by Ellis et al. and by Isgur, give rise to a nonzero charge radius, which has the right sign and size if the hyperfine interactions account for Δ-N splitting. This is easy to see intuitively - the Δ is heavier than the nucleon - therefore the hyperfine interactions must be repulsive between quarks with parallel spin. But this leads to a net repulsion between the two down quarks in the neutron (which have parallel spins) and to a negative charge radius - as observed experimentally.

Other consequences of mixing a 70-plet into the ground state baryons are the violation of the Moorhouse selection rule which forbids the photo decay of ^4P excited protons into the proton. With a 70 plet component in the proton these decays are allowed and the amplitude and helicity structure are in reasonable agreement with experiment. There is a similar selection rule by Faiman and Plane involving the $\bar{K}N$ decay of ^4P strangeness minus one hyperons whose observed violation is in reasonable agreement with the computation based on 70 plet mixing in the nucleon as argued by Koniuk et al.

Finally one can actually compute the decay amplitudes of a large set of excited nucleons, say the n=1 and n=2 configurations. This involves very laborious computations, which were recently completed by Isgur and Koniuk. A very noteworthy feature of these computations is that many states do decouple from the elastic channel as a result of the mixings due to hyperfine interactions. These states do account for so called 'missing' states in the quark model - which were not found in partial wave analyses of elastic scattering. Thus hyperfine interactions help tremendously in the comparison between quark model computations and experimental data.

REFERENCES

For a more complete set of references see:

Isgur, Nathan, lectures at the XVIth International School of Subnuclear Physics, Erice, Italy (1978) to appear.

Karl, Gabriel, Proceedings of the XIXth International Conference on High Energy Physics Tokyo (1978), edited by S. Homma, M. Kawaguchi and H. Miyazawa (Phys. Soc. of Japan, Tokyo 1979) p. 135.

Hey, A.J.G., Lecture at the European Physical Society Meeting in Geneva 1979 (Southampton preprint).

More recent work:

SU_6 Breaking: N. Isgur et al. Phys. Rev. Letters 41, 1269 (1978).

Baryon Decays: R. Koniuk and N. Isgur, Toronto preprints (1979).

Ground State Baryons: N. Isgur and G. Karl, Phys. Rev. D20, 1191 (1979).

Charmed Baryons: L.A. Copley et al. Phys. Rev. D19, 768 (1979).

Inelastic Electron-Proton and γ-Proton Scattering and the Structure of the Nucleon*

J. D. Bjorken and E. A. Paschos
Stanford Linear Accelerator Center, Stanford University, Stanford, California 94305

(Received 10 April 1969)

A model for highly inelastic electron-nucleon scattering at high energies is studied and compared with existing data. This model envisages the proton to be composed of pointlike constituents ("partons") from which the electron scatters incoherently. We propose that the model be tested by observing γ rays scattered inelastically in a similar way from the nucleon. The magnitude of this inelastic Compton-scattering cross section can be predicted from existing electron-scattering data, indicating that the experiment is feasible, but difficult, at presently available energies.

I. INTRODUCTION

ONE of the most interesting results emerging from the study of inelastic lepton-hadron scattering at high energies and large momentum transfers is the possibility of obtaining detailed information about the structure, and about any fundamental constituents, of hadrons. We discuss here an intuitive but powerful model, in which the nucleon is built of fundamental pointlike constituents. The important feature of this model, as developed by Feynman, is its emphasis on the infinite-momentum frame of reference.

It is argued that when the inelastic scattering process is viewed from this frame, the proper motion of the constituents of the proton is slowed down by the relativistic time dilatation, and the proton charge distribution is Lorentz-contracted as well. Then, under appropriate experimental conditions, the incident lepton scatters instantaneously and incoherently from the individual constituents of the proton, assuming such a concept makes sense.

We were greatly motivated in this investigation by Feynman, who put the above ideas into a highly workable form. In Sec. II, we discuss the basic ideas and equations for the model as they apply to electron-proton scattering. Two models are then discussed in detail, with interesting consequences for the ratio of electron-proton and electron-neutron scattering. For a broad class of such models, we find a sum rule which indicates that, although it is not difficult to fit the data within $\sim 50\%$, it is more difficult to do better; the observed cross section is uncomfortably small.

In Sec. III, we look for stringent tests of Feynman's picture. We propose that, under similar experimental conditions, inelastic Compton scattering can also be calculated within the model. It is shown that the ratio of inelastic electron-proton to inelastic γ-proton scattering, under identical kinematical conditions, is model-independent and of order unity, provided the proton constituents (which Feynman calls "partons") possess unit charge and spin 0 or ½. We propose experiments which can measure inelastic Compton scattering. To this end we have estimated the yield and background curves, and we conclude that such experiments may be feasible at energies available to SLAC.

II. INELASTIC e-p SCATTERING

The basic idea in the model is to represent the inelastic scattering as quasifree scattering from pointlike constituents within the proton, when viewed from a frame in which the proton has infinite momentum. The electron-proton center-of-mass frame is, at high energies, a good approximation of such a frame. In the infinite-momentum frame, the proton is Lorentz-contracted into a thin pancake, and the lepton scatters instantaneously. Furthermore, the proper motion of the constituents, of partons, within the proton is slowed down by time dilatation. We can estimate the interaction time and the lifetime of the virtual states within the proton. By using the notation of Fig. 1, we find the following.

Time of interaction:

$$\tau \approx 1/q^0 = 4P/(2M\nu - Q^2)\,, \tag{2.1}$$

where q^0 was calculated in the lepton-proton center-of-mass frame.

Lifetime of virtual states:

$$T = \{[(xP)^2+\mu_1{}^2]^{1/2} + [(1-x)^2P^2+\mu_2{}^2]^{1/2} - [P^2+M_p{}^2]^{1/2}\}^{-1} = \frac{2P}{(\mu_1{}^2+p_{1\perp}{}^2)/x+(\mu_2{}^2+p_{2\perp}{}^2)/(1-x)-M_p{}^2}\,. \tag{2.2}$$

If we now require that

$$\tau \ll T\,, \tag{2.3}$$

then we can consider the partons, contained in the proton, as free during the interaction. Furthermore, if we consider large momentum transfers $-q^2 \gg M^2$, then we expect the scattering from the individual partons to be incoherent. The above conditions appear to be satisfied in the high-energy, large-momentum-transfer experiments at SLAC.

The kinematics for e-p inelastic scattering have been discussed in many places, in as many different nota-

* Work supported by the U. S. Atomic Energy Commission.

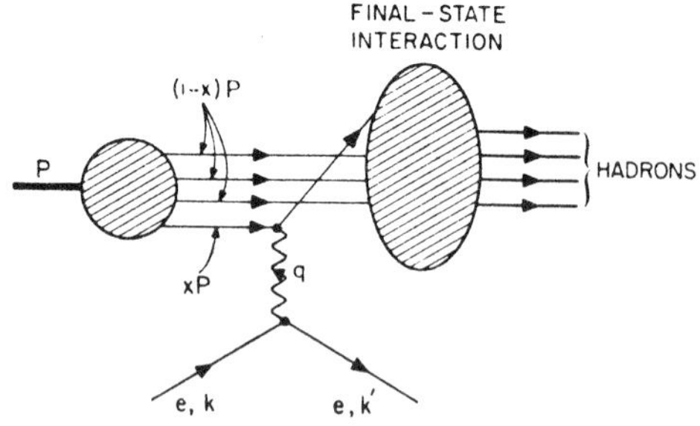

FIG. 1. Kinematics of lepton-nucleon scattering in the parton model.

tions.[1] We recall here that

$$\frac{EE'}{\pi}\frac{d\sigma}{dq^2d\nu}=\frac{d\sigma}{d\Omega dE'}=\frac{\alpha^2}{4E^2\sin^4(\frac{1}{2}\theta)}\times[W_2\cos^2(\tfrac{1}{2}\theta)+2W_1\sin^2(\tfrac{1}{2}\theta)], \quad (2.4)$$

where W_1 and W_2 are functions of the two invariants

$$\nu=(E-E')=q\cdot P/M,$$
$$Q^2=-q^2=4EE'\sin^2(\tfrac{1}{2}\theta), \quad (2.5)$$

evaluated in the laboratory frame. The ratio W_1/W_2 is bounded. Using

$$\frac{W_1}{W_2}=\left(1+\frac{\nu^2}{Q^2}\right)\frac{\sigma_t}{\sigma_t+\sigma_l}, \quad (2.6)$$

and the approximation $\frac{1}{2}\theta\ll 1$, we can write

$$\frac{d\sigma}{d\Omega dE'}\cong\frac{\alpha^2}{4E^2\sin^4(\frac{1}{2}\theta)}W_2(q^2,\nu)\left[1+\left(\frac{\sigma_t}{\sigma_t+\sigma_l}\right)\frac{\nu^2}{2EE'}\right], \quad (2.7)$$

where σ_t and σ_l (≥ 0) are the absorption cross sections for transverse and longitudinal photons.[2]

It remains to calculate the invariant functions W_1 and W_2, especially W_2. Within the model, the virtual photon interacts with one of the partons, while the rest remain undisturbed during the interaction. The interaction with the parton is as if the parton were a free, structureless particle. The cross section $d\sigma/d\Omega dE'$ is then a sum over individual electron-parton interactions appropriately weighted by the parton charge and momentum. For a free particle of any spin and unit charge, elementary calculation yields

$$W_2(\nu,q^2)=\delta(\nu-Q^2/2M)=M\delta(q\cdot P-\tfrac{1}{2}Q^2), \quad (2.8)$$

while for W_1, we have

$$\sigma_t=0 \quad \text{for spin } 0,$$
$$\sigma_l=0 \quad \text{for spin } \tfrac{1}{2}, \quad (2.9)$$

with an indeterminate result for higher spins.

[1] S. Drell and J. Walecka, Ann. Phys. (N. Y.) **28**, 18 (1964); J. D. Bjorken, Phys. Rev. **179**, 1547 (1969).

[2] L. Hand, in *Proceedings of the Third International Symposium on Electron and Photon Interaction at High Energies, Stanford Linear Accelerator Center, 1967* (Clearing House of Federal Scientific and Technical Information, Washington, D. C., 1968).

At infinite momentum, we visualize the intermediate state from which the electron scatters as follows:

(a) It consists of a certain number N of free partons (with probability P_N).

(b) The longitudinal momentum of the ith parton is a fraction x_i of the total momentum of the proton:

$$\mathbf{p}_i=x_i\mathbf{P}. \quad (2.10)$$

(c) The mass of the parton, before and after the collision, is small (or does not significantly change).

(d) The transverse momentum of the parton before the collision can be neglected, in comparison with $\sqrt{(Q^2)}$, the transverse momentum imparted as $p\to\infty$.

With these assumptions, it should be a good approximation to write, at infinite momentum,

$$\mathbf{p}_i^\mu\cong x_i\mathbf{P}^\mu. \quad (2.11)$$

The question of corrections to this approximation has been studied by Drell, Levy, and Yan.[3] We shall not consider them further here.

The contribution to W_2 from a single parton of momentum $x\mathbf{P}^\mu$ and charge Q_i is then

$$W_2^{(i)}=x_iQ_i^2M\delta(q\cdot x_iP-\tfrac{1}{2}Q^2)$$
$$=Q_i^2M\delta\left(q\cdot P-\frac{Q^2}{2x_i}\right)=Q_i^2\delta\left(\nu-\frac{Q^2}{2Mx_i}\right). \quad (2.12)$$

The factor x_i in front is necessary to ensure that

$$\lim_{E\to\infty}\frac{d\sigma^{(i)}}{dq^2}=Q_i^2\left(\frac{4\pi\alpha^2}{q^4}\right), \quad (2.13)$$

consistent with the Rutherford formula. For a general distribution of partons in the proton, we have

$$W_2(\nu,q^2)=\sum_N P(N)\langle\sum_i Q_i^2\rangle_N\times\int_0^1 dx\, f_N(x)\delta\left(\nu-\frac{Q^2}{2xM}\right). \quad (2.14)$$

Here $P(N)$ is the probability of finding a configuration of N partons in the proton, $\langle\sum_i Q_i^2\rangle_N$ equals the average value of $\sum_i Q_i^2$ in such configurations, and $f_N(x)$ is the probability of finding in such configurations a parton with longitudinal fraction x of the proton's momentum, that is, with four-momentum xP^μ.

Upon integrating over x, we find

$$\nu W_2(\nu,q^2)=\sum_N P(N)\langle\sum_i Q_i^2\rangle_N xf_N(x)\equiv F(x), \quad (2.15)$$

with

$$x=Q^2/2M\nu. \quad (2.16)$$

Therefore νW_2 is predicted to be a function of a *single*

[3] S. D. Drell, D. J. Levy, and T.-M. Yan (private communication).

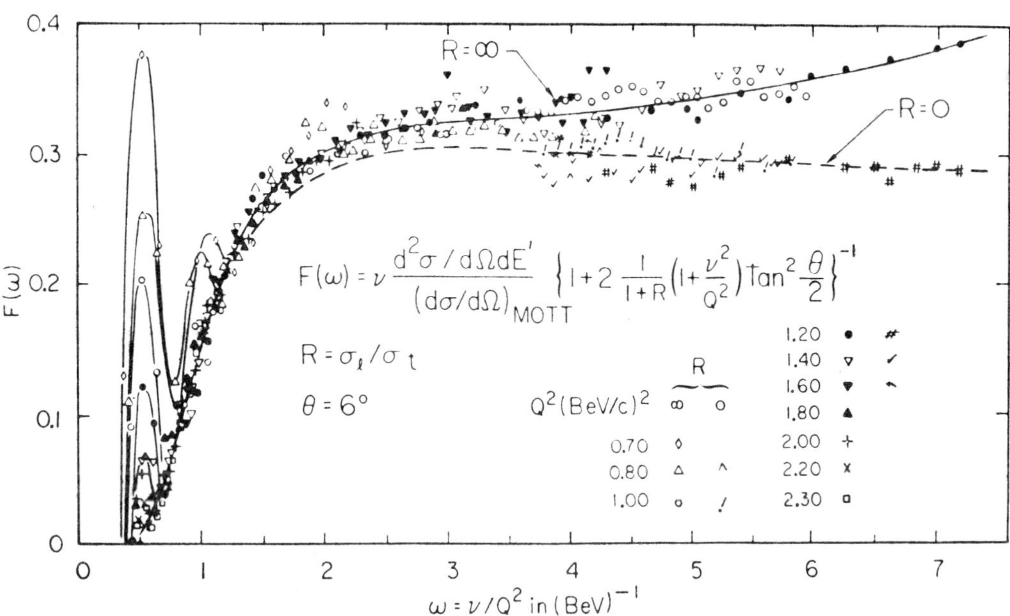

FIG. 2. Plot of the data as a function of ν/Q^2.

variable ν/Q^2, a feature apparently satisfied by the data.[4] Furthermore, the model provides an interpretation of the nature of the function $F(x)/x$: It is the mean square of the charge of partons with four-momentum xP^μ. The experimental[4] determination of $F(x)$ is shown in Fig. 2.

Before going into detailed models for $f_N(x)$, we notice that

$$f_N(x_1)=\int dx_2\cdots dx_N f_N(x_1,\cdots,x_N)\delta(1-\sum_i x_i),$$

$$\int_0^1 dx_1 f_N(x_1)=1, \tag{2.17}$$

where $f_N(x_1,\cdots,x_N)$ is the joint probability of finding partons (irrespective of charge) with longitudinal fractions $x_1,\cdots,x_N$. It follows that f_N is a symmetric function of its arguments. Therefore,

$$\int_0^1 x_1 dx_1 f_N(x_1)=\frac{1}{N}\int dx_1\cdots dx_N(\sum_i x_i)\times f_N(x_1,\cdots,x_N)\delta(1-\sum_i x_i)=1/N. \tag{2.18}$$

Putting together (2.18) and (2.15), we obtain a sum rule

$$\int dx\, F(x)=\sum_N P(N)\langle\sum_i Q_i^2\rangle_N/N = \text{mean-square charge per parton}. \tag{2.19}$$

[4] See the rapporteur talk of W. H. K. Panofsky, in *Proceedings of the Fourteenth International Conference on High-Energy Physics, Vienna, 1968* (CERN, Geneva, 1968), pp. 36–37, based on the work of E. Bloom, D. Coward, H. DeStaebler, J. Drees, J. Litt, G. Miller, L. Mo, R. Taylor, M. Breidenbach, J. Friedman, G. Hartmann, H. Kendall, and S. Loken.

Numerically,

$$\frac{Q^2}{2M}\int\frac{d\nu}{\nu}W_2=\int dx\, F(x)\approx 0.16, \tag{2.20}$$

yielding a rather small mean-square charge per parton.

In the following models, out of ignorance, we shall choose one-dimensional phase space for the distribution function $f_N(x_1\cdots x_N)$; that is,

$$f_N(x_1\cdots x_N)=\text{const}. \tag{2.21}$$

An elementary calculation yields

$$f_N(x)=(N-1)(1-x)^{N-2}. \tag{2.22}$$

A. Three-Quark Model

Assuming that the proton is made up of three quarks with the usual charges,[5] we obtain

$$\nu W_2=f_3(x)=2x(1-x),\quad x=Q^2/2M\nu. \tag{2.23}$$

While the data support $\nu W_2\to$ const as $\nu\to\infty$ (or $x\to 0$), the model predicts that νW_2 should vanish, a result not dependent on the specific choice of $f_3(x)$, but only on the fact that f_3 is normalizable. In fact, within our one-dimensional model, if the number of partons is held finite, then the cross section vanishes as $x\to 0$. If, and only if,

$$\lim_{N^2\to\infty} N^2 P(N)=\text{const}\neq 0 \tag{2.24}$$

will νW_2 approach a constant as $x\to 0$ ($\nu/Q^2\to\infty$). This is shown in the Appendix.

[5] M. Gell-Mann, Phys. Letters 8, 214 (1964); C. Zweig, CERN Report Nos. TH 401, 402, 1964 (unpublished).

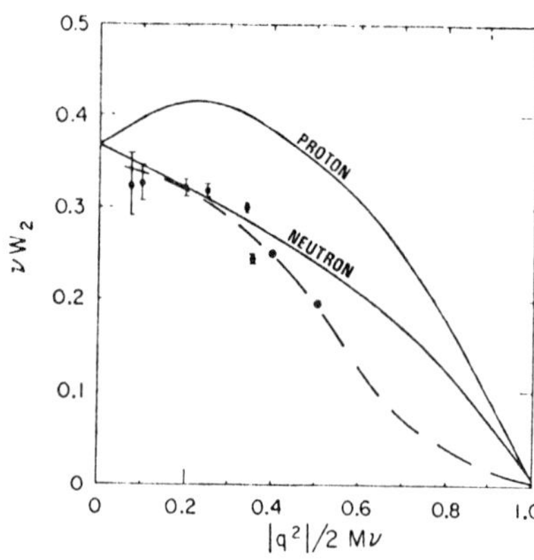

FIG. 3. Plot of the results for a model of three quarks in a sea of quark-antiquark pairs. The dashed line is visual fit through the experimental points of Ref. 4.

B. Three Quarks in a Background of Quark-Antiquark Pairs

In order to try to improve the model, we assume that in addition to the three quarks there is a distribution of quark-antiquark pairs (the "pion cloud"?). The mean-square charge of the cloud we take to be statistical

$$\langle \sum Q_i^2 \rangle_{\rm cloud}/N = \tfrac{1}{3}[(\tfrac{2}{3})^2+(\tfrac{1}{3})^2+(\tfrac{1}{3})^2]=2/9. \quad (2.25)$$

Therefore,

$$\langle \sum_i Q_i^2 \rangle_N = 1+2/9(N-3)=2/9N+\tfrac{1}{3} \quad \text{for the proton}$$
$$=\tfrac{2}{3}+2/9(N-3)=2/9N \quad \text{for the neutron.} \quad (2.26)$$

For $P(N)$, we choose

$$P(N)=C/N(N-1), \quad (2.27)$$

on the grounds that it is simple and has the asymptotic behavior which makes $\nu W_2 \to$ const as $x\to 0$. Before we begin, we note, from (2.26) and (2.19), that

$$\int_0^1 dx\, F(x)=\int_0^1 dx\, \nu W_2 = 2/9+\tfrac{1}{3}\langle 1/N\rangle > 0.22 \text{ (proton)}$$
$$=2/9=0.22 \text{ (neutron)}$$
$$\approx 0.16 \text{ (expt.)}. \quad (2.28)$$

Thus we cannot expect a fit better than $\sim 50\%$ to the data. Inserting (2.27), (2.26), and (2.22) into the expression (2.15) for $F(x)$, we can perform the sum. For the proton,

$$F(x)=x\sum_{N=3,5\cdots} P(N)(N-1)(1-x)^{N-2}(2/9N+\tfrac{1}{3})$$
$$=C\sum_{N=3,5\cdots}\left(\frac{2}{9}+\frac{1}{3N}\right)x(1-x)^{N-2}$$
$$=C\left\{\frac{2}{9}\frac{(1-x)}{(2-x)}+\frac{1}{6}\frac{x}{(1-x)^2}\left[\ln\left(\frac{2-x}{x}\right)-2(1-x)\right]\right\}. \quad (2.29)$$

The normalization constant C is determined from the condition

$$1=\sum_{N=3,5\cdots} P_N(x)=C\sum_{N=3,5\cdots}\frac{1}{N(N-1)}=C(1-\ln 2). \quad (2.30)$$

For the neutron, the term in square brackets is omitted; thus the final expressions are

$$F_p(x)=\frac{1}{1-\ln 2}\left\{\frac{2}{9}\frac{(1-x)}{(2-x)}+\frac{1}{6}\frac{x}{(1-x)^2}\times\left[\ln\left(\frac{2-x}{x}\right)-2(1-x)\right]\right\}, \quad (2.31)$$

$$F_n(x)=\frac{2}{9}\left(\frac{1}{1-\ln 2}\right)\left(\frac{1-x}{2-x}\right).$$

It is clear that the results are model-dependent and not to be taken too seriously. There is a need for a model-independent check of the basic assumptions. In Sec. III we discuss the corresponding process with electrons replaced by γ rays as a test of the basic idea of the calculation.

In Fig. 3, we compare Eqs. (2.31) with the experiment. The shape of the curve is in fair agreement, and could be improved by suppressing the contribution of three-parton configurations. On the other hand, the over-all normalization is off, as discussed below [(2.28)]. Readjustment of the coefficients P_N or the distributions $f_N(x)$ of longitudinal fraction cannot improve this feature. The ratio of neutron cross section to proton remains nearly constant and about 0.8 over a large range of x, although the ratio approaches 1 as $x\to 0$.

According to (2.9), if the partons all have spin $\frac{1}{2}$, we expect $\sigma_l/\sigma_t \to 0$; if they are spinless, we find instead $\sigma_t/\sigma_l\to 0$. A finite ratio would indicate that both kinds are present.

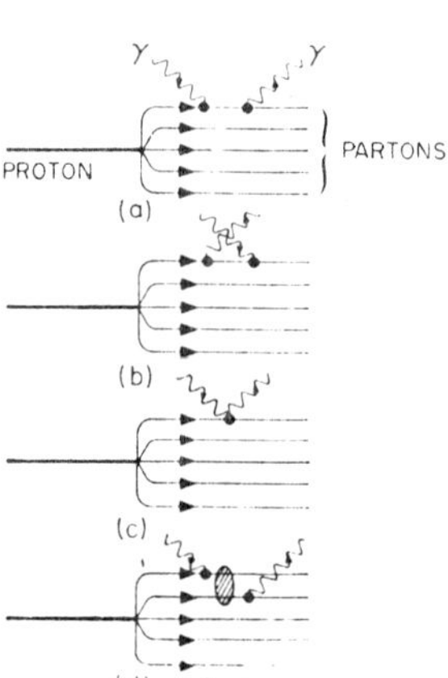

FIG. 4. Diagrams contributing to inelastic Compton scattering.

III. INELASTIC COMPTON SCATTERING

In this section, we suggest that a good way to find out about the internal structure of the proton is to look at it. That is, one should measure the inelastic scattering of photons from protons, yielding a photon plus anything else. The yield can be predicted in terms of this model. We visualize the inelastic scattering again as incoherent scattering from the partons in the proton according to the point cross section. The elementary photon-parton scattering goes by the diagrams shown in Fig. 4. Diagram (c) in Fig. 4, does not occur for spin-$\frac{1}{2}$ partons. Its contribution is evidently very similar to the case of electron scattering. We argue that exchange-terms such as in diagrams (d) and (e) can be ignored on the basis that the lifetime of the intermediate states, between absorption and emission of the photon, is of order $E_{\text{c.m.}}^{-1}$ in the γ-proton center-of-mass system and much less than the lifetime of the virtual-parton states of the proton estimated in (2.2). Furthermore, because we require the momentum transfer between the photons to be large (>1 BeV/c), the parton will necessarily have transverse momentum much larger than the average momentum within the proton and the probability that it interacts with another parton within the proton at such high $P_\perp$ is very small.

The kinematics for the process is illustrated in Fig. 5, with the value

$$\begin{aligned} s&=(k+p)^2\approx 2k\cdot p=(2Mk)_{\text{lab}}\gg M^2,\\ t&=(k-k')^2=-2k\cdot k'=-[4kk'\sin^2(\tfrac{1}{2}\theta)]_{\text{lab}},\\ M\nu&=(k-k')\cdot p=M(k-k')_{\text{lab}},\\ u&\approx -s-t. \end{aligned}\tag{3.1}$$

We require $-t$ to be large, as we already mentioned, so that the process is incoherent, and only Figs. 4(a)–4(c) (for integer spin) need be taken into account. The method for calculation is the same as in Sec. II: We take the Compton cross section from point partons and average over the parton momentum distributions and proton configurations. This elementary photon-parton interaction is given (for spin-$\frac{1}{2}$ partons) by the Klein-Nishina formula written in terms of the Mandelstam variables s, t, u. If the parton carries the full momentum $P(x=1)$, we have

$$\begin{aligned}\frac{d\sigma}{dtd\nu}&=\frac{4\pi\alpha^2}{s^2}\delta\left(\nu+\frac{t}{2M}\right)Q_i^4 \quad \text{for spin 0}\\ &=\frac{-2\pi\alpha^2}{s^2}\left(\frac{s}{u}+\frac{u}{s}\right)\delta\left(\nu+\frac{t}{2M}\right)Q_i^4 \quad \text{for spin } \tfrac{1}{2},\end{aligned}\tag{3.2}$$

which can be combined into the form

$$\frac{d\sigma}{dtd\nu}=\frac{4\pi\alpha^2}{s^2}\left(1-R\frac{(s+u)^2}{2su}\right)\delta\left(\nu+\frac{t}{2M}\right)Q_i^4,\tag{3.3}$$

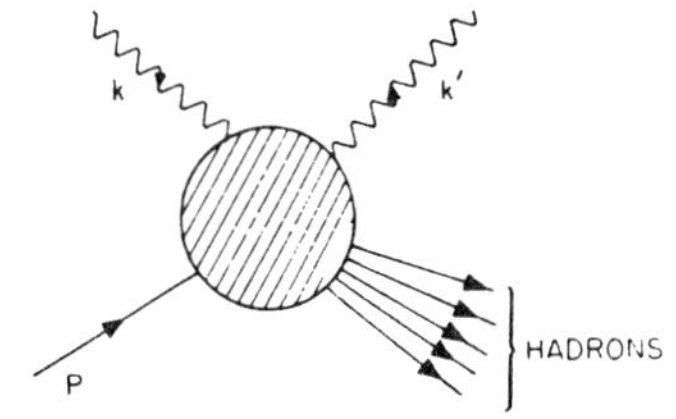

FIG. 5. Kinematics of Compton scattering.

with

$$\begin{aligned} R&=1 \quad \text{for spin } \tfrac{1}{2}\\ &=0 \quad \text{for spin } 0.\end{aligned}\tag{3.4}$$

If the parton has longitudinal fraction x, we make the replacements (we are here in the infinite-momentum frame)

$$s\to xs,\quad u\to xu,\quad t\to t,\quad \delta(\nu+t/2M)\to\delta(\nu+t/2Mx).\tag{3.5}$$

We then multiply by the distribution function $f_N(x)$ and by $P(N)$, integrate over x, and sum over N [cf. Eqs. (2.12)–(2.15)] to obtain

$$\frac{d\sigma}{dtd\nu}=\sum_N\int_0^1 dx\,P(N)f_N(x)\frac{4\pi\alpha^2}{x^2s^2}\times\left[1-R\frac{(s+u)^2}{2su}\right]\delta\left(\nu+\frac{t}{2Mx}\right)\langle\sum_i Q_i^4\rangle_N.\tag{3.6}$$

Integrating over x and expressing the result in laboratory variables, Eq. (3.1) gives the final result:

$$\frac{d\sigma}{d\Omega dk'}=\frac{\alpha^2}{4k'\sin^4(\frac{1}{2}\theta)}\frac{\nu}{kk'}\left[1+R\frac{\nu^2}{2kk'}\right]\times\sum_N P(N)xf_N(x)\langle\sum_i Q_i^4\rangle_N,\tag{3.7}$$

with

$$x=-t/2M\nu.\tag{3.8}$$

Making the identification

$$k\leftrightarrow E,\quad k'\leftrightarrow E',\quad -t\leftrightarrow Q^2,\quad R\leftrightarrow\sigma_t/(\sigma_t+\sigma_l),\tag{3.9}$$

we find a remarkable correspondence between (3.7) and (2.14), (2.9), and (2.7), the corresponding cross section for electrons. The only changes are the additional factor ν^2/kk' and the replacement $Q_i^2\to Q_i^4$. Even the factor in square brackets dependent on parton spin is the same. We conclude that within the validity of this parton model, for partons of unit charge ($Q^2=Q^4$) and spin 0 or $\frac{1}{2}$, the ratio of electron scattering to γ scattering is a model-independent number of order

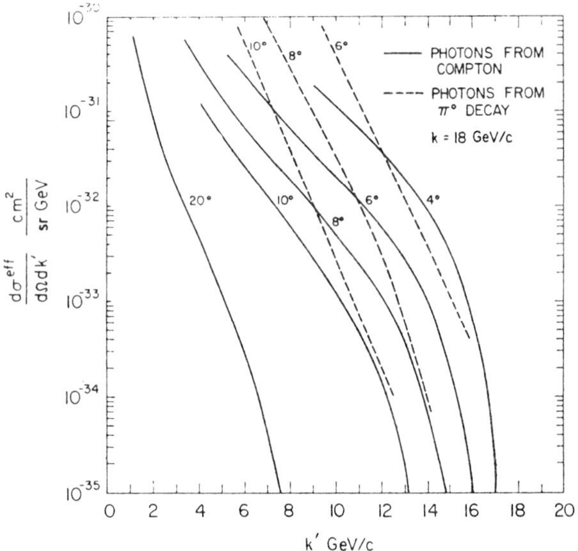

FIG. 6. Double-differential laboratory cross section for inelastic Compton scattering for an 18-GeV incident bremsstrahlung spectrum. The solid curve corresponds to the signal. The dotted curve is the background of γ's from π^0's using the data of Ref. 8 as discussed in the text.

unity. In general,[6]

$$\left(\frac{d\sigma}{d\Omega dE'}\right)_{\gamma p}=\frac{\nu^2}{kk'}\left(\frac{d\sigma}{d\Omega dE'}\right)_{ep}\langle\sum_i Q_i^4\rangle/\langle\sum_i Q_i^2\rangle, \quad (3.10)$$

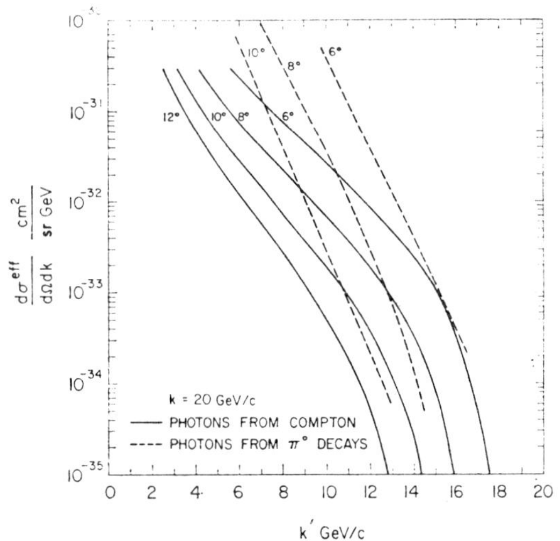

FIG. 7. Inelastic Compton scattering for a 20-GeV incident bremsstrahlung spectrum. The background curves are the same as in Fig. 6.

[6] We point out that the argument used in Sec. II in deriving the sum rule can be repeated here, giving

$$Q^2\int\left(\frac{d\sigma}{d\Omega dE}\right)_{\gamma p}\frac{kk'}{\nu^2\sigma_R}d\nu=\sum_N P(N)\langle\textstyle\sum_i Q_i^4\rangle_N/N,$$

where $\sigma_R=[\alpha^2/4k^2\sin^4(\frac{1}{2}\theta)](1+R\nu^2/2kk')$. We have also assumed the same momentum distribution $f_N(x)$ for the spin-0 and spin-$\frac{1}{2}$ partons.

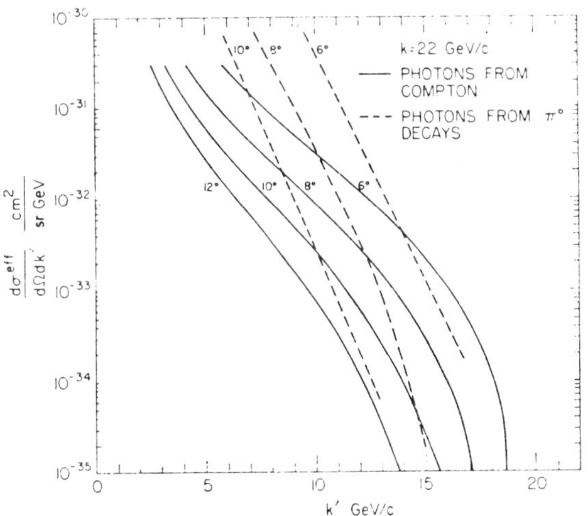

FIG. 8. Inelastic Compton scattering for a 22-GeV incident bremsstrahlung spectrum. The background curves are the same as in Fig. 6.

where for any operator $O(N)$ we have

$$\langle O\rangle=\sum_N P(N)O(N)f_N(x). \quad (3.11)$$

For our model of three quarks in a cloud of quark-antiquark pairs, there exist upper and lower limits for $\langle\sum Q^4\rangle/\langle\sum Q^2\rangle$. We note from (2.25) and (2.26), for the proton, that

$$\langle\sum_i Q_i^4\rangle_N=\frac{11}{27}+\frac{2}{27}(N-3)=\frac{1}{3}\langle\sum_i Q_i^2\rangle_N+\frac{2}{27}$$
$$=\frac{5}{9}\langle\sum_i Q_i^2\rangle_N-\frac{4}{81}N. \quad (3.12)$$

Therefore, for identical kinematical regions[7] we have

$$\frac{1}{3}\frac{\nu^2}{EE'}\left(\frac{d\sigma}{d\Omega dE'}\right)_{ep}<\left(\frac{d\sigma}{d\Omega dE'}\right)_{\gamma p}<\frac{5}{9}\frac{\nu^2}{EE'}\left(\frac{d\sigma}{d\Omega dE'}\right)_{ep}. \quad (3.13)$$

In principle it should be possible, if the model is correct, to distinguish between fractional and integer-charged partons.

IV. BACKGROUND AND RATE ESTIMATES FOR INELASTIC γ-RAY EXPERIMENT

We consider an experiment in which a bremsstrahlung beam is sent through hydrogen and the inelastically scattered γ ray is detected with momentum k' at angle θ. The main problem, in principle, is to differentiate the Compton γ rays from the γ rays coming from the decays of photoproduced π^0's. Here, we calculate the effective

[7] These bounds depend on the implicit assumptions of Eq. (2.25). A smaller lower bound of $\frac{1}{9}$ is obtained if we assume that all the pairs are made up of charges $\frac{1}{3}$ and $-\frac{1}{3}$.

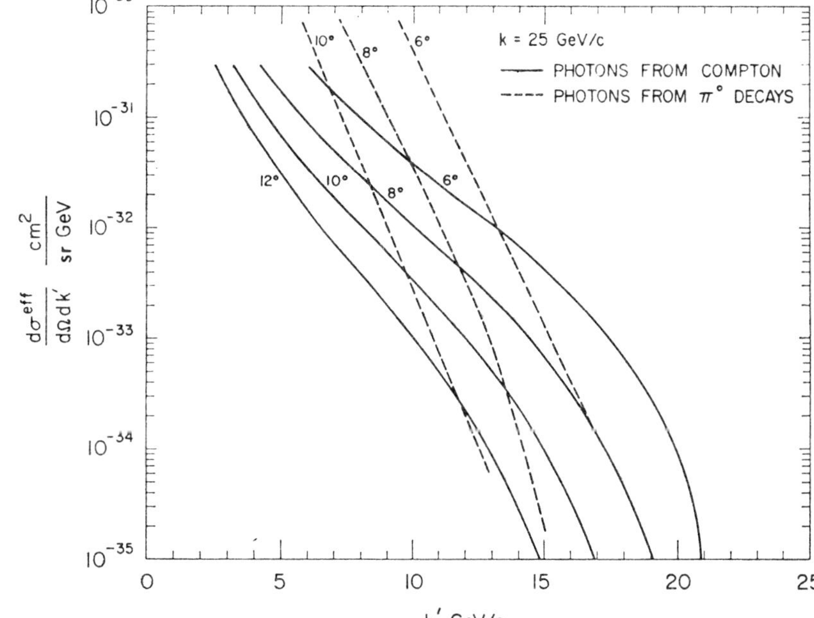

FIG. 9. Inelastic Compton scattering for a 25-GeV incident bremsstrahlung spectrum. The background curves are the same as in Fig. 6.

cross section for the production of Compton-scattered γ rays by folding the Compton cross section over the bremsstrahlung spectrum. We then calculate for comparison the corresponding background γ rays, estimated at SLAC from the beam survey experiments[8] for charged π's. We define

$$\frac{d\sigma^{\text{eff}}}{d\Omega dk'}=\int_{k'+|t|/2M}^{E\text{ electron}}\frac{dk}{k}\frac{d\sigma^{\gamma p}}{d\Omega dk'}. \tag{4.1}$$

We assume, optimistically, that the partons have unit charge and spin $\frac{1}{2}$; from (3.3) and (4.1) we obtain

$$\frac{d\sigma^{\text{eff}}}{d\Omega dk'}=\frac{4\alpha^2}{M^2k'}\int_{1/2}^{M(E-k')/4Ek'\sin^2(\frac{1}{2}\theta)}\lambda d\lambda F(\lambda)$$

$$\times\left[1-\frac{4k'\lambda\sin^2(\frac{1}{2}\theta)}{M}+\frac{8k'^2\lambda^2\sin^4(\frac{1}{2}\theta)}{M^2}\right], \tag{4.2}$$

with

$$\lambda=M\nu/Q^2. \tag{4.3}$$

We have calculated this expression for several incident electron energies as a function of k' and θ, using for $F(\lambda)$ the values given in Ref. 2. The results are shown as Figs. 6–9. In the same figures are shown our estimates of the corresponding background from the decay of photoproduced π^0's into γ rays. The estimate was made by assuming that the yield of π^+ measured in the SLAC beam survey experiment[8] equals the π^0 yield. We thereby obtain, for the effective cross section per nucleon,

$$\left(\frac{d\sigma_\pi}{d\Omega dk'}\right)_{\text{eff}}=(\text{yield})_{\pi^+}\bigg/0.7\int_{t=0}^{0.3}tdt\text{ (g/rad length of Be)}$$

$$\times(\text{Avogadro No.})$$

$$=8.2\times10^{-25}(\text{yield})^{\pi^+}\text{ cm}^2/\text{sr BeV}. \tag{4.4}$$

The terms in the denominator have the following origin: The thin-target bremsstrahlung spectrum is tdk/k, where t is the thickness in radiation lengths (r.l.) (the target was 0.3 r.l. Be). The factor 0.7 is a thick-target correction calculated by Tsai and Van Whitis.[9] (Yield)$^{\pi^+}$ is taken from the SLAC User's Handbook[8] and the γ-ray flux is obtained by folding the π^0-decay spectrum into (4.4):

$$\left(\frac{d\sigma}{d\Omega dk_\gamma'}\right)^\gamma=2\int_{k_\gamma}^{E}\left(\frac{d\sigma_\pi}{d\Omega dk'}\right)_{\text{eff}}\frac{dk'}{k'}\approx\frac{2}{k_\gamma}\int_{k_\gamma}^{\infty}\left(\frac{d\sigma_\pi}{d\Omega dk'}\right)_{\text{eff}}dk'$$

$$\approx(2E_0/k_\gamma\theta)(8.2\times10^{-25})(\text{yield})^{\pi^+}$$

$$(\text{as function of }k_\gamma), \tag{4.5}$$

where in the last step we have used the empirical observation that

$$d\sigma_\pi/d\Omega dk'\sim e^{-k'\theta/E_0}, \tag{4.6}$$

with

$$E_0\approx0.154\text{ BeV}.$$

In Figs. 6–9, the background is that from an 18-BeV bremsstrahlung beam. It is expected that this background increases slowly with beam energy, and keeps

[8] SLAC User's Handbook, Sec. D.1, Figs. 1 and 2 (unpublished).

[9] Y. S. Tsai and Van Whitis, Phys. Rev. **149**, 1948 (1966).

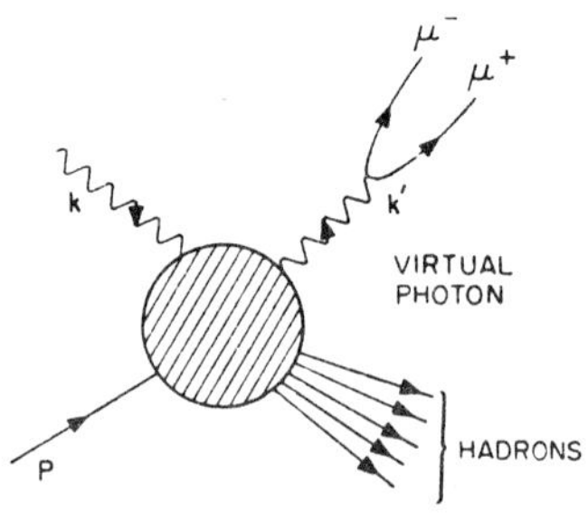

FIG. 10. Muon-pair production by inelastic Compton scattering.

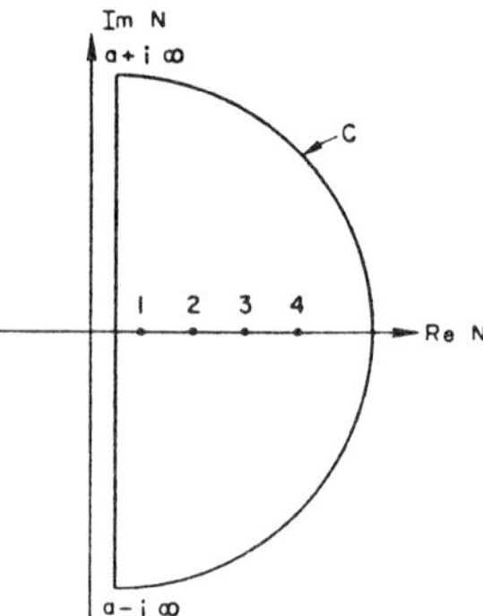

FIG. 11. Contour of integration for the Sommerfeld-Watson transform.

the same shape, in particular, the exponential dependence on transverse momentum. However, as the primary electron energy increases, the γ-ray spectrum from the inelastic Compton process, if it exists, is displaced upward in energy, so that for 20-BeV electrons and above, there is a region where the Compton signal dominates the π^0 noise. In any case, it will certainly be necessary to compare the γ-ray spectrum with that of the π^+ under the same conditions as reassurance that Compton γ rays are indeed being seen.

A variation of this experiment is to consider the inelastic Compton terms in μ-pair photoproduction (see Fig. 10). We have not analyzed this process in detail. The rate is diminished by a factor roughly[10] $\sim(2\alpha/3\pi)[\ln(E_{\max}/m_\mu)-3.5]\sim 1/350$ but if the charged pions are absorbed immediately downstream from the target, the background muon flux from $\pi^\pm$ decay can be reduced by a factor $\sim 1/700$ as well. Furthermore, the two muons are strongly correlated in angle, providing a quite unique signature. All this is encouragement that perhaps the background is manageable. The "singles" background from the Bethe-Heitler diagrams, for which the undetected muon predominantly goes in the forward direction, is interesting, as well, and very likely exceeds the singles rate from Compton μ pairs. But this is also of interest in testing μ-e universality at very high q^2. The "Bethe-Heitler" muons probably dominate the background muons from π^+ decay, but the necessary estimates have not yet been made.

APPENDIX

We attain a condition between $P(N)$ and $\langle\sum Q_i^2\rangle_N$ which guarantees that $F(x)$ is analytic at $x=0$ and equal to a nonzero constant. Using the Sommerfeld-Watson transformation, we rewrite (2.15) as an integral over the contour shown in Fig. 11:

$$F(x)=2x\int_c NP(2N+1)\langle\sum_i Q_i^2\rangle_{(2N+1)}\frac{(1-x)^{2N-1}e^{i\pi N}}{\sin\pi N}dN. \tag{A1}$$

The contribution of the semicircle at infinity is negligible. For what remains in the integral, we can use (for $x\to 0$)

$$x(1-x)^{2N-1}\approx xe^{-(2N-1)x}=-\tfrac{1}{2}\partial e^{-(2N-1)x}/\partial N, \tag{A2}$$

and integrate by parts:

$$F(x)=\int_{a-i\infty}^{a+i\infty} e^{-(2N-1)x}\frac{\partial}{\partial N}\times\left[\frac{NP(2N+1)\langle\sum_i Q_i^2\rangle_{(2N+1)}e^{i\pi N}}{\sin\pi N}\right]dN. \tag{A3}$$

In the limit of $x\to 0$, we have

$$F(x)\longrightarrow \left.\frac{NP(2N+1)\langle\sum_i Q_i^2\rangle_{(2N+1)}e^{i\pi N}}{\sin\pi N}\right|_{a-i\infty}^{a+i\infty} \xrightarrow[N\to a-i\infty]{} NP(2N+1)\langle\sum_i Q_i^2\rangle_{(2N+1)}. \tag{A4}$$

Therefore $F(x)$ approaches a constant if and only if $P(N)\to c/N\langle\sum Q_i^2\rangle_N$. For $\langle\sum_i Q_i^2\rangle_N$ linear in N, this reduces to the condition mentioned in the text.

ACKNOWLEDGMENTS

We thank R. P. Feynman, J. Weyers, and our colleagues at SLAC for many helpful conversations.

[10] R. H. Dalitz, Proc. Phys. Soc. (London) A64, 667 (1951).

VOLUME 25, NUMBER 5 PHYSICAL REVIEW LETTERS 3 AUGUST 1970

MASSIVE LEPTON-PAIR PRODUCTION IN HADRON-HADRON COLLISIONS AT HIGH ENERGIES*

Sidney D. Drell and Tung-Mow Yan
Stanford Linear Accelerator Center, Stanford University, Stanford, California 94305
(Received 25 May 1970)

On the basis of a parton model studied earlier we consider the production process of large-mass lepton pairs from hadron-hadron inelastic collisions in the limiting region, $s\to\infty$, Q^2/s finite, Q^2 and s being the squared invariant masses of the lepton pair and the two initial hadrons, respectively. General scaling properties and connections with deep inelastic electron scattering are discussed. In particular, a rapidly decreasing cross section as $Q^2/s\to 1$ is predicted as a consequence of the observed rapid falloff of the inelastic scattering structure function νW_2 near threshold.

Feynman's parton model[1] for deep-inelastic weak or electromagnetic processes is an expression of the impulse approximation as applied to elementary-particle interactions. In order to apply the impulse approximation we demand the following. We analyze the bound system – be it a nucleon or nucleus – in terms of its constitutents, called "partons." Nucleons are the "partons" of the nucleus and the "partons" of a nucleon itself are still to be deciphered. If we specify the kinematics so that the partons can be treated as instantaneously free during the sudden pulse carrying the large energy transfer from the projectile (or lepton) then we can neglect their binding effects during the interaction and we can treat the kinematics of the collision as between two free particles, the projectile and the parton. Moreover, if we are in a kinematic regime so that energy is approximately conserved along with momentum across the interaction vertex of the parton with the weak or electromagnetic current, the conditions for applying the impulse approximation are satisfied.

The Bjorken limiting region[2] satisfies this condition for the deep inelastic electron scattering from protons as viewed from a certain class of $P\to\infty$ or infinite-momentum frames. The "partons" constituting a proton are strongly bound together as viewed in the rest frame. However, if their bound state can be formed primarily by momentum components that are limited in magnitude below some fixed maximum – i.e., if there exists a finite $k_{\max}$ – then as viewed in an infinite-momentum frame these parton states are long-lived by virtue of the characteristic time dilatation. The derivation of this intuitively appealing picture from a canonical quantum field, modified by imposing a maximum constraint on $k_\perp$, has been discussed as well as its applicability to the particular class of amplitudes with "good currents."[3] In particular, the ratio $Q^2/2M\nu$, where $Q^2>0$ is the negative of the square of the invariant momentum transfer and $q\cdot P=M\nu$, measures the fraction $x\equiv Q^2/2M\nu$ of the longitudinal momentum on the parton from which the electron scatters and is a finite fraction $0<x<1$ in the Bjorken limit.

It is easy to show that the ratio x must be finite in order to apply the impulse approximation. Otherwise as x approaches very close to 0 or 1 we will be forced to deal with very slow partons in the $P\to\infty$ system, or, as seen in the rest system of the proton, with the high-momentum extremities of the bound-state structure, and for these the impulse approximation breaks down.

The beauty of the electron scattering is that it allows us to "tune" the mass of the virtual photon line as we choose to probe finite x. However when we return to the world of only real external hadrons, we have no large mass since $Q^2\to M^2$ while $2M\nu\to s$, the total collision energy. In this case x becomes very small,[1] or "wee." Our condition for applying the impulse approximation also fails and the value of the parton con-

VOLUME 25, NUMBER 5 PHYSICAL REVIEW LETTERS 3 AUGUST 1970

cept is less certain.[4] The impulse approximation also applies to electron-positron pair annihilation into a specific hadron H plus anything else: $e^+ + e^- \to H +$ "anything" in the deep-inelastic region of large lepton-pair mass squared q^2 and large invariant energy transfer ν. In an infinite-momentum frame of the detected hadron, this process can be described as the creation of an essentially free parton-antiparton pair and its subsequent decay into final states.

If we want to find other processes which satisfy the kinematical constraints allowing application of the impulse approximation we need look for interactions at high energies s which absorb or produce a lepton system of huge mass Q^2 such that the ratio Q^2/s is finite. An observable class of processes meeting this requirement is production of massive lepton pairs in hadron-hadron collisions,[5] viz.,

$$p + p \to (\mu^+ \mu^-) + \cdots . \tag{1}$$

Our remarks apply equally to any colliding pair such as (pp), $(\bar{p}p)$, (πp), (γp) and to final leptons $(\mu^+\mu^-)$, $(e\bar{e})$, $(\mu\nu)$, and $(e\nu)$.

What is going on here can be best illustrated in a center-of-mass frame. If a massive state with $Q^2 \sim s$ emerges from one of the colliding protons A or B as in Fig. 1(a), it is impossible to satisfy both energy and momentum conservation in the overall collision and at the same time exchange only "wee" partons between A and B.[6] Hence this process will not be related directly to the total nucleon-nucleon cross section[7] in which, as discussed by Feynman, it is the "wee" partons with $x \sim 1$ GeV$/\sqrt{s}$ that cannot tell "right" from "left" in Fig. 1(a) that are responsible for σ_T. In contrast, the dominant amplitude in (1) in a model of the nucleon with a finite momentum $k_{\max}$ in its ground-state structure will be the production

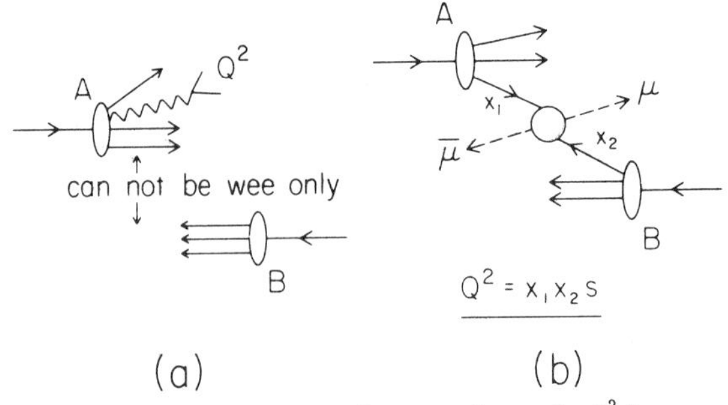

FIG. 1. (a) Production of a massive pair Q^2 from one of the hadrons in a high-energy collision. In this case it is kinematically impossible to exchange "wee" partons only. (b) Production of a massive pair by parton-antiparton annihilation.

of the massive lepton pair by annihilation of an antiparton-parton pair as illustrated in Fig. 1(b). Viewed from the center-of-mass frame a hard (i.e., non-"wee") parton moving to the right, say, annihilates on a similar antiparton headed to the left and the resulting system is very massive since their energies add whereas their momenta subtract. It is easy to show that if a pair of mass Q^2 is formed,

$$Q^2 = x_1 x_2 s, \quad 0 < x_{1,2} < 1, \tag{2}$$

where $x_{1,2}$ are the fractions of the longitudinal momenta of their respective hadrons carried by the annihilating parton pair. Clearly for finite Q^2/s one is here dealing with hard partons and with the same region of momenta as probed by deep-inelastic scattering experiments which measure the parton distribution in $x = Q^2/2M\nu$. In this process we are measuring over a range of their values as constrained by (2) for fixed Q^2/s.

We now turn to a calculation of (1) in the deep-inelastic region of finite $\tau = Q^2/s$ with $s \to \infty$. The general expression for the cross section is

$$\frac{d\sigma}{dQ^2} = \left(\frac{4\pi\alpha^2}{Q^2}\right)\left(1 - \frac{4m^2}{Q^2}\right)^{1/2}\left(1 + \frac{2m^2}{Q^2}\right)\{[s-(M_1+M_2)^2][s-(M_1-M_2)^2]\}^{-1/2} W(Q^2, s), \tag{3}$$

where a spin average is understood and

$$\begin{aligned} W(Q^2, s) &\equiv -16\pi^2 E_1 E_2 \int (dq)\delta(q^2-Q^2)\int (dx) e^{-iqx}\langle P_1 P_2^{(\mathrm{in})}|J_\mu(x)J^\mu(0)|P_2 P_1^{(\mathrm{in})}\rangle \\ &= -16\pi^2 E_1 E_2 \int (dq)\delta(q^2-Q^2)\textstyle\sum_n (2\pi)^4\delta^4(P_1+P_2-q-P_n)\langle P_1 P_2^{(\mathrm{in})}|J_\mu|n\rangle\langle n|J^\mu|P_2 P_1^{(\mathrm{in})}\rangle. \end{aligned} \tag{4}$$

In (4) E_1, $\vec{P}_1$, M_1 and E_2, $\vec{P}_2$, M_2 are the energies, momenta, and masses of the two initial hadrons and m is the muon mass. Since we will directly imitate the steps in our preceding analyses of deep inelastic processes,[3] we first define a true infinite-momentum frame by boosting from the collision center-of-mass frame by a velocity $\beta/(1-\beta^2)^{1/2} = 2P/\sqrt{s}$ in a direction orthogonal to the collision axis. The four-vector momenta of the two incident colliding hadrons are then, for $s \gg M^2$,

$$P_1^{\mu} = \left(P + \frac{s}{8P}, \tfrac{1}{2}\sqrt{s}, 0, P\right), \quad P_2^{\mu} = \left(P + \frac{s}{8P}, -\tfrac{1}{2}\sqrt{s}, 0, P\right). \tag{5}$$

VOLUME 25, NUMBER 5 PHYSICAL REVIEW LETTERS 3 AUGUST 1970

We can now let $P\to\infty$ for large but finite s: $P \gg \sqrt{s} \gg M$. The energy in the collision is represented by a transverse momentum mismatch of the colliding hadrons. For a parton, or a baryon or meson quantum in our field-theory model, to be exchanged between them without introducing an asymptotically large momentum transverse to either of the two hadron lines, the parton momentum is restricted to a fraction $\sim M/\sqrt{s}$ along the $\vec{P}$ axis and to a finite value $\sim M$ orthogonal to it. This constraint corresponds to the "wee" parton condition in the center-of-mass frame of the colliding hadrons. In the $P\to\infty$ frame (5) this constraint satisfies the condition of finite transverse momentum imposed on our field-theory model.

In this frame[8] we can repeat steps developed in earlier work of undressing the current operator by the U matrix: $J_\mu(0)=U^{-1}j_\mu(0)U$ where $j_\mu(0)$ is the current operator expressed in terms of free fields. Furthermore the energy differences between the eigenstate $|P_1P_2{}^{(\mathrm{in})}\rangle$ and the components of $U|P_1P_2{}^{(\mathrm{in})}\rangle$ can be ignored in the limit $s\to\infty$ for Q^2/s finite; the same is true for $|n\rangle$ and $U|n\rangle$. This is so because the invariant mass of the individual system of particles moving along $\vec{P}_1$ and $\vec{P}_2$, respectively, in (5), or to the right and left in the center-of-mass frame, is finite as a result of the transverse momentum cutoff imposed. This mass is thus negligible compared with the invariant mass $\frac{1}{2}\sqrt{s}$ appearing in (5). In other words the impulse approximation is good and energy as well as momentum is conserved across the electromagnetic current vertex in (4). This leads to the simplification of (4) in the Bjorken limit for $P\to\infty$, $s \gg M^2$, Q^2/s finite, to

$$\lim{}_{\mathrm{Bj}} W = -16\pi^2 E_1E_2\int(dq)\delta(q^2-Q^2)\int(dx)\,e^{-iqx}\langle U(P_1P_2)^{\mathrm{in}}|j_\mu(x)j^\mu(0)|U(P_2P_1)^{\mathrm{in}}\rangle \tag{6}$$

and in our model, as described in earlier work, to a factorization of the U matrix:

$$|U(P_1P_2)^{\mathrm{in}}\rangle = |UP_1\rangle|UP_2\rangle. \tag{7}$$

Proceeding in analogy with II, Eqs. (72)-(78), we find for the annihilation of a boson pair (the same result obtains for a fermion pair with spin averaging)

$$(-)\int(dq)\delta(q^2-Q^2)\int(dx)\,e^{-iqx}\langle k_1k_2|j^\mu(x)j_\mu(0)|k_2k_1\rangle=(2\pi)^4\delta(Q^2-(k_1+k_2)^2)\frac{-\lambda^2}{(2\pi)^6(2\omega_1)(2\omega_2)}(k_1-k_2)_\mu(k_1-k_2)^\mu$$

$$=\frac{\lambda^2}{16\pi^2E_1E_2}\delta(x_1x_2-\tau);\quad \tau\equiv Q^2/s<1,$$

where λ^2 is the square of the charge of an individual parton and we have used the high-energy approximation for the dominant large components of the momenta $k_1{}^\mu = x_1P_1{}^\mu$, $k_2{}^\mu=x_2P_2{}^\mu$. Inserting the identity

$$1=\int_0^1 d\left(\frac{1}{\omega_1}\right)\int_0^1 d\left(\frac{1}{\omega_2}\right)\delta\left(x_1-\frac{1}{\omega_1}\right)\delta\left(x_2-\frac{1}{\omega_2}\right),$$

$$\lim{}_{\mathrm{Bj}}W(Q^2,s)\equiv\frac{1}{\tau}\mathfrak{F}(\tau)=\sum_a\lambda_a{}^2\int_0^1 d\left(\frac{1}{\omega_1}\right)\int_0^1 d\left(\frac{1}{\omega_2}\right)\delta\left(\frac{1}{\omega_1\omega_2}-\tau\right)$$

$$\times\left\langle UP_1\left|\delta\left(x_{1,a}-\frac{1}{\omega_1}\right)\right|UP_1\right\rangle\left\langle UP_2\left|\delta\left(x_{2,\bar a}-\frac{1}{\omega_2}\right)\right|UP_2\right\rangle, \tag{8}$$

where the summation over types of partons with charges λ_a pairs a parton of type a in $|UP_1\rangle$ with its antiparton $\bar a$ in $|UP_2\rangle$ and vice versa. By comparison with (78), (79), and (80) of II we see that (8) can be written as

$$\mathfrak{F}(\tau)=\sum_a(\lambda_a)^{-2}\int_1^\infty d\omega_1\int_1^\infty d\omega_2\delta(\omega_1\omega_2-1/\tau)F_{2a}(\omega_1)F_{2\bar a}{}'(\omega_2) \tag{9}$$

in terms of the invariant structure functions $F_{2a}(\omega_1)=\nu W_2{}^{(\lambda_a)}$ introduced in the deep-inelastic scattering analyses [see (78) of II] for ω_1 times the probability of finding parton of type a in the proton (or hadron A) with a momentum fraction $x_1=1/\omega_1$. $F_{2\bar a}{}'(\omega_2)$ has the same significance for the corresponding antiparton distribution in hadron (B).

The differential cross section (3) now assumes the simple form in the scaling limit

$$\frac{d\sigma}{dQ^2}=\left(\frac{4\pi\alpha^2}{3Q^2}\right)\left(\frac{1}{Q^2}\right)\mathfrak{F}(\tau)=\left(\frac{4\pi\alpha^2}{3Q^2}\right)\left(\frac{1}{Q^2}\right)\int_0^1dx_1\int_0^1dx_2\delta(x_1x_2-\tau)\sum_a\lambda_a{}^{-2}F_{2a}(x_1)F_{2\bar a}{}'(x_2), \tag{10}$$

VOLUME 25, NUMBER 5 PHYSICAL REVIEW LETTERS 3 AUGUST 1970

where we have rewritten the invariant structure functions in terms of momentum fraction x. $4\pi\alpha^2/3Q^2$ is just the total cross section for $e\bar{e}$ annihilation into (point) muon pairs in the relativistic limit.

Equation (10) is the central result of this Letter and is a formal expression of our earlier discussion. We conclude with several remarks about general features of this result.

(1) The observed[9] rapid decrease of the inelastic structure functions $F_2(x) = \nu W_2$ as $x \to 1$ leads in (2) and (10) to a prediction of a very rapid falloff in $\mathfrak{F}(\tau)$ with increasing $\tau = Q^2/s$. If we assume that the parton and antiparton have identical momentum distributions in the proton and this is common for all parton types λ, we can compute $d\sigma/dQ^2$ directly from measured $F_2(x)$, finding a very rapid falloff in the cross section as shown in Fig. 2, even though the model consists of pointlike constituents. This is in qualitative accord with preliminary experimental findings.[5] However, a quantitative comparison with data requires a more detailed discussion about the kinematic cuts in momenta and angles of the leptons involved in the experimental measurements. This will be done in a forthcoming paper.

(2) The angular distribution of the vector $\vec{q} \equiv \vec{p}_+ + \vec{p}_-$, the total momentum of the muon pair, is peaked along the incident nucleon's direction in the lab system. This follows from the observation that $q \cdot P_1 = (x_1 P_1 + x_2 P_2) \cdot P_1 \cong \frac{1}{2} x_2 s$ is an invariant and in terms of laboratory variables $q \cdot P_1 \cong E_1 q^0 (1-\cos\theta)$, with $M_2 E_1 \cong \frac{1}{2}s$, so that $1-\cos\theta \sim O(1/q^0)$.

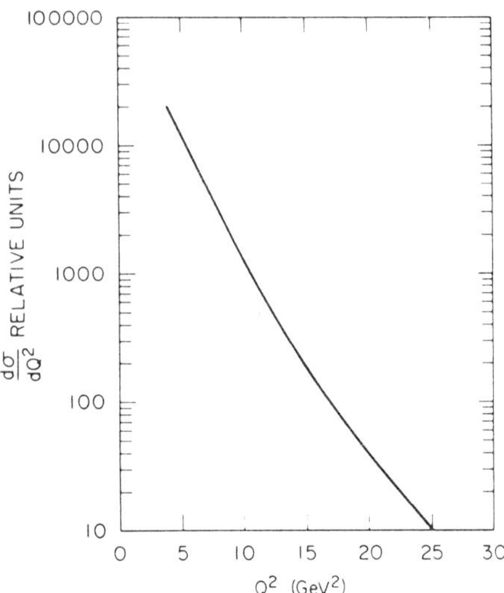

FIG. 2. $d\sigma/dQ^2$ computed from Eq. (10) assuming identical parton and antiparton momentum distributions and with relative normalization.

(3) The virtual photon will be predominantly transversely polarized if it is formed by annihilation of spin-$\frac{1}{2}$ parton-antiparton pairs. This means a distribution in the di-muon rest system varying as $(1+\cos^2\theta)$ rather than $\sin^2\theta$ as found in Sakurai's[10] vector-dominance model, where θ is the angle of the muon with respect to the timelike photon momentum. The model used in Fig. 2 assumed identical parton-antiparton distributions and hence the spin-$\frac{1}{2}$ partons play the predominant role as in the scattering experiments.[9]

(4) The full range of processes of the type (1) with incident p, $\bar{p}$, π, K, γ, etc., affords the interesting possibility of comparing their parton and antiparton structures. (In particular no relation between the parton and antiparton spectra need be assumed, as we did in Fig. 2, for an initial $\bar{p}p$ state.) Not only are the variations important but so are the cross-section magnitudes as measures of effective λ's.

(5) The factoring in (7) is possible only because "wee" parton exchanges are absent in our model for processes with hard partons to which an impulse approximation applies. This would not be the case if our theoretical model were enlarged to include a "wee" region of prominence (perhaps due to neutral vector exchanges). Presumably such quanta are needed to generate Feynman's spectrum of "wee" or infrared quanta, dx/x for explaining real hadron cross sections.[11] Since the impulse condition does not apply in these interactions we cannot compute purely hadronic processes by our techniques as in (6). However we can ask what implications there will be for our results for massive lepton-pair production if such "wee" quanta are introduced and modify (7) by initial-state interactions.

For example, suppose we include the "wee" parton exchanges between the two systems A and B before or after the parton-antiparton annihilation takes place. Precisely because the transferred momenta are "wee," these interactions can change the invariant mass of individual groups A and B in Fig. 1 only by a finite amount and the fractions of their longitudinal momenta by order of 1 GeV$/\sqrt{s}$. These corrections therefore do not affect our arguments leading to (6) which in turn implies (2) and the general scaling (9). Therefore although the invariant function $\mathfrak{F}(\tau)$ will be modified from (9) by the "wee" exchanges, the general scaling property will not be affected. Based on this observation we would

VOLUME 25, NUMBER 5 PHYSICAL REVIEW LETTERS 3 AUGUST 1970

like to emphasize that although "wee" exchanges must survive at infinite energies to account for a nonvanishing total cross section of hadron-hadron collisions, they are not relevant to the Bjorken scaling behavior of deep-inelastic lepton processes such as electron scattering, electron-positron annihilation, and the massive muon-pair production in proton-proton scattering considered here. A nontrivial Bjorken scaling behavior and the validity of the impulse approximation for these processes are independent of whether or not the total cross section for hadrons vanishes at high energies.

Note added in proof.—We have just received a preprint of a related study by Altarelli, Brandt, and Preparata at Rockefeller University, who have made an analysis based on a direct study of the light cone behavior of (4). The factor $(1/Q^2)^2$ multiplying the scaling function $\mathfrak{F}(\tau)$ in (10) is replaced in their work by one of form $[g(\tau)+bh(\tau)/Q^2]$, where b is a parameter to be fitted. The main physical difference between their approach and ours lies in the following: Our model predicts scaling behavior at asymptotic energies and therefore parameters with a dimension such as Regge trajectories do not enter explicitly. We have argued that the scaling law (10) is a result of the validity of the impulse approximation which, in turn, follows from the fact that we are dealing with hard partons in the asymptotic region $s\to\infty$ with $\tau\equiv Q^2/s$ fixed. [For more details on this point, see S. D. Drell, Stanford Linear Accelerator Center Report No. SLAC-PUB-745, 1970 (unpublished), and S. D. Drell and T. M. Yan (to be published).] On the other hand, the work of Altarelli, Brandt, and Preparata does not satisfy scaling, and Regge parameters appear explicitly. In parton language, this means that in their model, based on a light cone analysis, the "wee" parton region is directly responsible. A test of the two approaches will be possible when the s as well as Q^2 dependence of the cross section is known.

---

*Work supported by the U. S. Atomic Energy Commission.

[1]R. P. Feynman, Phys. Rev. Lett. 23, 1415 (1969), and unpublished talks.

[2]J. D. Bjorken, Phys. Rev. 179, 1547 (1969).

[3]S. D. Drell, D. J. Levy, and T. M. Yan, Phys. Rev. Lett. 22, 744 (1969), and Phys. Rev. 187, 2159 (1969), and Phys. Rev. D 1, 1617 (1970); T. M. Yan and S. D. Drell, *ibid.* 1, 2402 (1970). The last four papers will be referred to as Papers I, II, III, and IV, respectively.

[4]However, as suggested by Feynman, we may hope for clues to the behavior here by studying the deep-inelastic electron scattering as x decreases to very small values and thus be led to insights into what is going on here in the "wee" region.

[5]J. H. Christenson, G. S. Hicks, L. M. Lederman, P. J. Limon, B. G. Pope, and E. Zavattini, Bull. Amer. Phys. Soc. 15, 579 (1970). We thank Dr. Lederman for discussions of his experiment and its preliminary analysis.

[6]Simple calculation shows that the exchanged longitudinal momentum fraction is finite and given by $\frac{1}{2}[(\eta^2+4Q^2/s)^{1/2}-\eta]$, with η being the fractional longitudinal momentum of the virtual photon.

[7]For a different interpretation see S. Berman, D. Levy, and T. Neff, Phys. Rev. Lett. 23, 1363 (1969). For another approach using current commutators see R. F. Kogerler and R. M. Muradyan, Joint Institute for Nuclear Research Report No. E2-4791, 1969 (to be published). For a different application to photoproduction with a large momentum transfer as opposed to a large mass Q^2 see J. Bjorken and E. Paschos, Phys. Rev. 185, 1975 (1969).

[8]The following discussion also applies if we work in the center-of-mass frame of the two initial hadrons.

[9]E. Bloom *et al.*, Phys. Rev. Lett. 23, 930 (1969); M. Breidenbach *et al.*, *ibid.* 23, 935 (1969); R. Taylor, in *International Symposium on Electron and Photon Interactions at High Energies, Liverpool, England, September 1969*, edited by D. W. Braben (Daresbury Nuclear Physics Laboratory, Daresbury, Lancashire, England, 1970).

[10]J. J. Sakurai, Phys. Rev. Lett. 24, 968 (1970).

[11]We thank Dr. Ken Wilson for questions and stimulating discussions on this point.

Reprinted with permission from *Physics Reports,* **67** (1), pp. 109–121, S. Mandelstam, "General Introduction to Confinement."

General Introduction to Confinement

S. MANDELSTAM

Laboratoire de Physique Théorique de l'Ecole Normale Supérieure†

and

Department of Physics, University of California, Berkeley††, *U.S.A.*

1. Preliminaries

My aim in these lectures will be to give a brief survey of the general features of hadron structure and the confinement problem within the framework of quantum chromodynamics (Q.C.D.). I shall attempt to direct my presentation to the non-specialized audience present at this meeting.

When the quark model was first proposed, and especially when field theories of quarks began to be taken seriously, confinement appeared to be something of a mystery. It is no longer so. We understand the properties which the Q.C.D. vacuum must possess in order to confine, and we know of physical systems which possess analogues of such properties. Indeed, solid-state physicists have long been familiar with systems possessing confinement properties. The question whether the state of lowest energy of the system actually possesses such properties is more difficult. A number of approaches or approximation schemes indicates that the answer is yes but, at present, no explanation has received general support of particle theorists.

I shall divide my lectures into five parts:

1) General nature of hadronic states in confined systems.
2) Characterization of phases of pure Yang–Mills non-abelian systems. For simplicity we shall treat SU(2); the extension to SU(3) is straightforward.
3) Characterization of such phases for non-abelian systems with quarks.
4) The $1/N$ approximation, which simplifies the full Q.C.D. problem in many ways.
5) Survey of schemes for obtaining confinement and calculating hadronic masses.

While I shall mention direct points of contact with the lattice approach, my treatment will be within the framework of the continuum theory. Many of the features are parallel to those of lattice-gauge theory, which will be presented elsewhere at this meeting.

2. General nature of hadronic states in confined systems

An experimental feature, which has been of great help in understanding confinement, is that narrow hadronic resonances occur in linearly rising rotational sequences. For such sequences, the graph of J,

† Laboratoire Propre du C.N.R.S., associé à l'Ecole Normale Supérieure et à l'Université de Paris-Sud. Postal address: 24, rue Lhomond, 75231 Paris Cedex 05, France.

†† Permanent address.

the angular momentum, against s, the square of the centre-of-mass energy, is nearly a straight line. The experiments are easiest for baryons, and sequences extending as far as $J = 19/2$, or even higher, exist. Such sequences are intimately associated with confinement. If quarks were bound in a non-confining potential, the number of states would be finite or, for a Coulomb potential, the energy of the infinite sequence would approach a finite limit.

Rising rotational sequences are well described by a string model of hadrons, the dual-resonance model [1]. While the model is almost certainly qualitatively correct for high-angular-momentum states, it possesses defects which prevent its use, at any rate without modification, as a detailed theory of hadrons.

Nielsen and Olesen [2] suggested an interpretation of the strings of the dual-resonance model in terms of gauge field theory. Their idea is crucial to our understanding of confinement. Though we are mainly interested in a non-abelian theory, they took as their simplest model an abelian theory of a charged field interacting with a gauge field. The lagrangian was thus

$$\mathscr{L} = -\tfrac{1}{4}F_{\mu\nu}F^{\mu\nu} - \{(\partial_\mu + ieA_\mu)\Phi^*\}\{(\partial^\mu - ieA^\mu)\Phi\} + \mu^2\Phi^*\Phi - \lambda\{\Phi^*\Phi\}^2. \tag{1}$$

Note that the mass term has the wrong sign. As is well known to field theorists, the operator Φ then acquires a vacuum-expectation value,

$$\langle 0|\Phi|0\rangle \neq 0, \tag{2}$$

or, in the semi-classical model studied by Nielsen and Olesen,

$$|\Phi| = 0. \tag{3}$$

The conditions (2) or (3) are the conditions for superconductivity, and the introduction of a mass term of the wrong sign is the field-theorist's favourite trick for obtaining models with superconductivity. A super-conductor is a system where, in the ground state, the destruction operator for a charged particle has a non-zero expectation value. In other words, it is a coherent plasma of charged particles. In the system (1), the charged particle is elementary a "Higgs" particle. In metallic superconductors, it is a Cooper pair.

Nielsen and Olesen's suggestion was to identify the hadronic strings with the Landau–Ginsburg–Abrikosov vortices of quantized magnetic flux in the superconducting vacuum. Such vortices correspond to axially symmetric solutions of the semi-classical equations where, at large distances

$$\Phi = |\Phi| \exp\{i\phi\}, \qquad r = \surd(x^2 + y^2) \to \infty, \tag{4a}$$

ϕ being the azimuthal angle. If the vector potential has the behaviour

$$A_\phi = (er)^{-1}, \qquad r \to \infty \tag{4b}$$

the system will be a gauge transformation of the vacuum at infinity. The magnetic field F_{ij}, and the co-variant derivative of the charged field $(\partial_i - ieA_i)\Phi$, will both tend to zero.

Near the origin, however, there will be physical effects. Continuity requires that Φ have a zero at, at least, one point, which we take to be the origin. The vector potential must also depart from the form

(4b) in order to avoid the singularity at $r = 0$. We thus have the situation depicted in fig. 1. Within the vortex, both F_{12} and the co-variant derivative of Φ are non-zero, and physical energy is present.

If the radius a of the vortex is very small, the volume integral of F_{12}^2, and therefore the energy, will be very large. If a is very large, Φ will deviate from its optimum value over a large volume, and again the energy will be very large. It is thus plausible, and can easily be shown to be true, that there is some intermediate value of a which minimizes the energy, and that there is thus a solution of the classical field equations with the above boundary conditions.

The total flux within the vortex is easily calculated:

$$2\pi \int_0^\infty rF_{12}\,\mathrm{d}r = 2\pi \int_0^\infty r\frac{1}{r}\frac{\partial}{\partial r}(rA_\phi)$$

$$= 2\pi/e, \qquad \text{from (4b).} \tag{5}$$

A vortex must thus contain $2\pi/e$ units of magnetic flux. Outside the vortex the vacuum is in the superconducting state ($\Phi \neq 0$), and $F_{12} = 0$. Within the vortex, the vacuum is in the normal state ($\Phi = 0$), and $F_{12} = 0$.

The original Nielsen–Olesen vortices had to be endless; either infinite in length (as in the axially symmetric solution just described), or closed. The strings in the dual-resonance model are finite in length and open. As was pointed out by Nambu [3], the corresponding vortices would require monopoles at their ends to absorb the flux. We note that the Dirac quantization for monopole strength is precisely the flux quantization condition (5).

It now follows that we have a typical confinement situation of the type we expect in Q.C.D. A monopole–anti-monopole pair in a superconductor must be joined by a vortex of quantized magnetic flux. The energy of the flux tube will be proportional to the distance between the monopole and the anti-monopole; magnetic flux in a superconductor, unlike in a vacuum, cannot spread out. Thus *monopoles in a superconductor are confined.* A monopole and an anti-monopole in a superconductor, joined by a tube of magnetic flux, provides a mechanical model of a meson.

In Q.C.D. colour *charges*, not *monopoles*, are confined. This will occur if *electric*, rather than *magnetic* flux cannot spread out. (From now on we use the terms electric and magnetic to refer to their Q.C.D. analogues.) Such a suggestion was first made in its present context by 't Hooft. It is implied in the strong-coupling lattice model of Wilson, and the analogous hamiltonian model of Kogut and Susskind. The electric flux is forced along the links of the lattice.

An ordinary superconductor is a coherent superposition of charged objects. A vacuum which confines electric instead of magnetic flux may therefore be a coherent superposition of monopoles, as was suggested by Mandelstam [4] and 't Hooft [5]. Monopoles do not occur in continuum Q.E.D. unless

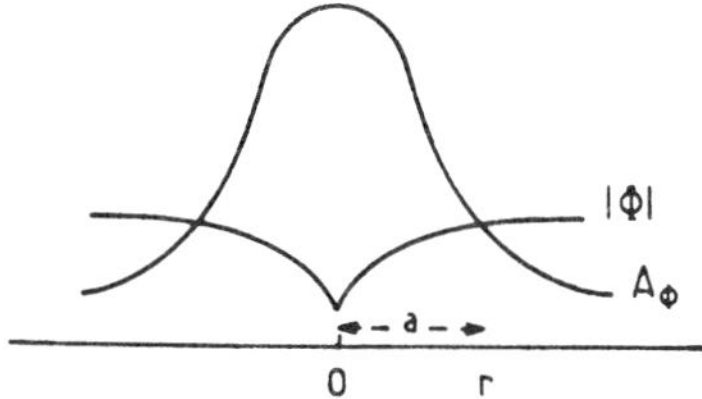

Fig. 1. Higgs field and vector potential in the Landau–Ginsburg–Abrikosov vortex.

they are introduced explicitly as elementary particles, and continuum Q.E.D. without monopoles cannot confine. In non-abelian gauge theories, on the other hand, monopoles can occur without explicit introduction. Yang and Wu [6] found a field distribution which may be called a monopole in the sense that there exists a gauge where

$$F_{ij} = \varepsilon_{ijk} r^k r^{-3}, \qquad r \to \infty. \tag{6}$$

't Hooft [7] showed that the Georgi–Glashow model, in which the non-abelian gauge symmetry is broken down to an abelian symmetry by a Higgs mechanism, admits of a monopole solution of the classical field equations.

The Yang–Mills vacuum may therefore be a coherent superposition of monopoles. Whether or not it is advantageous explicitly to construct the confining vacuum as such a superposition, one general feature thereof is almost certainly present in a confining vacuum. Since the magnetic field falls off slowly at large distances from a monopole, the amplitude of the virtual low-frequency oscillations of the magnetic field will be large compared to that of the perturbation-theory vacuum. I believe that this feature is common to all models of confinement.

3. Phases of non-abelian systems without quarks

In this section we shall discuss the characterization of the confining phase, and of other possible phases, of non-abelian gauge theories *without quarks*. By quarks we shall always understand particles in the fundamental representation of the gauge group; the spin is irrelevant for our purpose. We have already seen that the confinement property relates to the Yang–Mills system itself, though its most interesting implications concern systems with quarks. According to the $1/N$ approximation, to be discussed subsequently, quark couplings may be regarded as weak, and virtual processes involving quarks are unlikely to change qualitatively the nature of hadronic states.

A number of years ago, Wilson [8] proposed characterizing the confined vacuum by the operator

$$W = \mathrm{Tr}\!:\!\exp\left\{ \mathrm{i}g \int \mathrm{d}x^\mu \tau^\alpha A^\alpha_\mu \right\}\!:, \tag{7}$$

where the integral is around a large closed curve. The τ's are the matrices of the fundamental representation of the relevant group; the colons indicate that the exponential is to be expanded and the τ's ordered along the curve. The exponential in (7) is the analogue of the familiar abelian operator $\exp\{-\mathrm{i}e \int \mathrm{d}x^\mu A_\mu\}$. The untraced operator, taken along an open path, creates a tube of flux, along the path, of strength equal to one quark charge; it converts a quark and an anti-quark operator at the two ends into a gauge-invariant operator. By closing the curve we obtain an operator which creates a ring of flux.

We may obtain the Wilson confinement criterion by considering either time-like or fixed-time loops. Wilson himself considered time-like loops. If a heavy $\mathrm{Q\bar{Q}}$ pair is taken around the loop as in fig. 2, the loop integral will be the contribution of the quark–gluon interaction to the action. Without confinement the energy would approach a constant value as the quarks were separated. The action would therefore be proportional to the proper time of the quarks or, in other words, to the perimeter of the curve. In a confining vacuum the energy would be proportional to the distance between the quarks, and the action

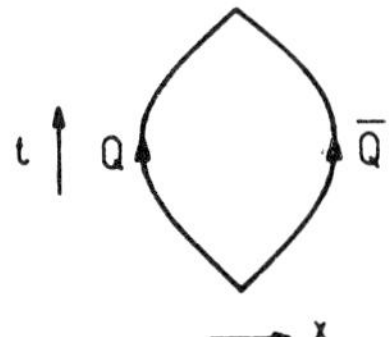

Fig. 2. Quarks taken round the Wilson path.

would be proportional to the area. Thus, since $\langle W\rangle$, evaluated in euclidean space-time, is equal to exp{-action},

$$\langle W\rangle = \exp\{-\text{perimeter}\} \equiv e^{-L}, \quad \text{no confinement}, \tag{8a}$$

$$\langle W\rangle = \exp\{-\text{area}\} \equiv e^{-A}, \quad \text{confinement}. \tag{8b}$$

Let us examine the process in more detail. If a $Q\bar{Q}$ pair traverses the path as in fig. 2, the quark–gluon interaction adds a factor (7) to the exponential of the action. Hence, when evaluating the vacuum-expectation value of any operator by functional integration, we should add such a factor. If we did not have confinement the process would only effect the vacuum-expectation values of operator in the neighbourhood of the loop itself whereas, in a confining vacuum, a tube of flux would move along the minimal area bounded by the loop. Hence

no confinement: $\langle 0|X(x)\, W|0\rangle = \langle 0|X(x)|0\rangle\, \langle 0|W|0\rangle$,
x distant from loop; (9a)

confinement: $\langle 0|X(x)\, W|0\rangle = \langle 0|X(x)|0\rangle\, \langle 0|W|0\rangle$,
x distant from minimal surface bounded by loop. (9b)

The operator X in (9) is any field operator, such as the hamiltonian or lagrangian density. If there are no massless particles, the equalities (9) will be approached exponentially as the distances are increased.†

We shall now obtain the Wilson condition by examining fixed-time loops. We have seen that the state $W|0\rangle$, i.e., a large flux loop, is more stable if we have confinement than if we do not. In the latter case it will simply disappear by diffusion, in the former it can only shrink to zero size – a process which takes a long time if the loop is large. The matrix element $\langle 0|W|0\rangle$, which would be zero if the state $W|0\rangle$ were completely stable, will therefore be smaller for the confining than for the non-confining vacuum. In any case, we should not expect the matrix element $\langle 0|W|0\rangle$ to be larger than e^{-L}, since the state $W|0\rangle$ is different from the vacuum over a volume proportional to L. For the confining vacuum, the matrix element would be smaller than this.

To obtain a more quantitative estimate of the matrix element with confinement, we consider a rectangular area. Suppose we regard two opposite sides as "particles" under the influence of the forces due to the other two sides. The "masses" of the "particles" will be Tl_1 where T is the energy per unit length of the flux tubes, i.e., the string tension. The distance between them will be l_2. Hence, from the

†I am indebted to J. Polchinski for a clarifying discussion regarding the foregoing argument.

familiar formula for the matrix element between well-separated one-particle states:

$$\langle 0|W|0\rangle = \exp\{-(Tl_1)\,l_2\} = \exp\{-TA\}. \tag{10}$$

Again we obtain the Wilson criterion.

The Wilson-loop operator creates a tube of electric flux. 't Hooft [9] suggested that one consider an operator M which creates a tube of *magnetic* flux. We recall that the vacuum is phase-rotated by 2π as we go round a Nielsen–Olesen vortex. In the corresponding non-abelian system, the vacuum would be gauge rotated by 2π. 't Hooft therefore considered an operator which is a pure gauge rotation except on the loop itself, and which gives a total rotation of 2π if one goes along a closed path which intersects the loop. Such an operator must necessarily introduce physical effects on the loop itself; in fact, it creates a vortex of magnetic flux. Mandelstam [10] has constructed explicit forms for such operators in both the abelian and non-abelian systems.

The above construction refers to a fixed time. A time-like 't Hooft loop would be the contribution to the action of a monopole anti-monopole pair taken around the path of fig. 2†.

't Hooft then suggested the following criterion for complete Higgs symmetry breaking, analogous to the Wilson criterion for confinement.

$$\langle M\rangle \to \exp\{-\text{perimeter}\}, \quad \text{no complete Higgs breaking}, \tag{11a}$$

$$\exp\{-\text{area}\}, \qquad \text{complete Higgs breaking}. \tag{11b}$$

The justification of this criterion would follow the justification of the Wilson criterion word for word, with magnetic and electric quantities interchanged. Eq. (11) is the *only* gauge-invariant, and therefore physically meaningful, definition of complete Higgs symmetry breaking. It is simply the Meissmer-effect definition of superconductivity.

't Hooft showed that the operators M and W possess simple fixed-time commutation relations, which have interesting consequences. We recall that, under a local gauge transformations, the open-path analogue of W transforms as follows:

$$W' \to \Omega^+(x)\, W'\, \Omega(y), \tag{12}$$

where x and y are the ends of the path. If W' is an almost-closed path which intersects the 't Hooft loop once and Ω is the gauge transformation associated with that loop, the rotations $\Omega(x)$ and $\Omega(y)$ differ by 2π. Thus, for SU(N)

$$\Omega(y) = e^{2\pi i/N}\Omega(x).$$

Hence, on taking the trace of W, we obtain the result

$$M^+ WM = e^{2\pi i/N}\, W,$$

or $$WM = e^{2\pi i/N} MW.$$

† For the reader familiar with the Dirac-strong formalism, an outline of the argument is as follows: if one attempts to calculate the vacuum-expectation value of $X(x)M$ by functional integration, the field configurations at $t+$ are rotated relative to those at $t-$ by Ω, the gauge rotation of the 't Hooft loop. By making a continuous gauge transformation, we can remove the discontinuities except on a sheet bounded by the loop, where we have a "Dirac sheet" of singularities. On rotating into a time-like direction, the result follows.

More generally, if the loops corresponding to M and W intersect n times (where n may be zero),

$$WM = e^{2\pi i n/N} MW \qquad \text{(fixed time)}. \tag{13}$$

If the perimeters of the loops touch one another, no simple commutation relations (13) hold.

The equations (13) lead to 't Hooft's argument that, *in a phase without massless particles*, either $\langle W\rangle$ or $\langle M\rangle$ must behave like $\exp\{-A\}$ when A is large. The equivalent of eq. (9) will be

$$\langle 0|WM|0\rangle = e^{2\pi i n_1/N} \langle 0|W|0\rangle \langle 0|M|0\rangle, \tag{14a}$$

$$\langle 0|MW|0\rangle = e^{2\pi i n_2/N} \langle 0|W|0\rangle \langle 0|M|0\rangle, \tag{14b}$$

under the conditions:

$\langle M\rangle, \quad \langle W\rangle \to e^{-L}$:	two perimeters distant,	(14c)
$\langle M\rangle \to e^{-L}, \quad \langle W\rangle \to e^{-A}$:	perimeter of M distant from minimal surface of W,	(14d)
$\langle M\rangle \to e^{-A}, \quad \langle W\rangle \to e^{-L}$:	perimeter of W distant from minimal surface of M.	(14e)

The extra phase factors in (14) are due to the fact that the gauge rotation can cause a phase change of $2\pi\, in/N$ in W, even if the paths are distant. In a system without massless particles, the fall-off will be exponential as the relevant distances are increased†.

We now observe that, by going into four euclidean dimensions, we can pass from the intersecting to the non-interacting case while keeping the perimeters of the two loops distant from one another. The value of n in (13) must change by one; on the other hand, the values of n_1 and n_2 in (14) cannot change if (14c) is satisfied. We thus arrive at a contradiction if both $\langle M\rangle$ and $\langle W\rangle$ behave like e^{-L}.

In a later paper, [11] 't Hooft showed that $\langle M\rangle$ and $\langle W\rangle$ cannot both behave like e^{-A}. If there are no massless particles we must have either electric or magnetic confinement but not both.

In all this work we notice a striking analogy between electric and magnetic operators. It may be asked whether non-abelian systems, like abelian systems, possess complete electric-magnetic duality. From the dynamical viewpoint the answer appears to be no. Kinematically the answer is yes; we have shown that one can construct dual vector potentials, at least with spatial smearing; gauge-invariant combinations of such variables constitute the physical variables [10]. The hamiltonian cannot be expressed in a simple form in terms of the dual vector potentials, however.

If we allow massless particles two further phases are possible. Thus, if we treat the SU(2) gauge theory for the sake of simplicity, we have the following four phases:

1) The "perturbation-theory" phase with real massless quarks and gluons. It is not known whether such a phase can exist; it probably cannot in an infra-red unstable system. If it can exist, it is not known whether it can be interpreted physically.
2) The "Georgi–Glashow" phase, where the SU(2) invariance is broken to U(1). This phase contains photons, electric charges and magnetic monopoles.

† The equations (14) are perhaps not an immediate consequence of the reasoning leading to eq. (9). For the gauge model without massless particles we can construct, i.e., the Higgs model, eqs. (14e) can be verified, in any order of perturbation theory, from the explicit formulas for M in ref. [10].

3) The phase with complete Higgs breaking, magnetic flux is combined in vortices; monopoles are combined.
4) The confinement phase. Electric flux is confined in vortices, charges are confined.

Phases 1) and 2) are self-dual, while 3) and 4) go into one another under electric-magnetic duality.

4. Phases of gauge theories with quarks

If actual quarks are present, the expectation value of the Wilson operator behaves like e^{-L}, even with confinement. Applying Wilson's argument, one observes that the Q and $\bar{Q}$ of fig. 2 could pick up a $\bar{Q}$ and a Q from the vacuum, and it no longer takes a large energy to separate them. In the fixed-time argument, a large loop of electric flux can now easily break up by the creation of $Q\bar{Q}$ pairs (fig. 3). Both arguments thus lead to a behavior e^{-L}.

The 't Hooft proof that either $\langle M\rangle$ or $\langle W\rangle$ must behave like e^{-A} no longer holds, since the operator M cannot be constructed in the manner described in the previous section. A rotation by 2π changes the phase of quark creation operators, and 't Hooft's construction would produce physical effects away from the loop. A more complicated construction is required for the operator M.

We thus require a new criterion for confinement. In fact, our ability to recognize confinement in nature depends either on the weakness of quark couplings or on the existence of (exactly or approximately) conserved quantum numbers external to Q.C.D. Included in the first category of experiments is the observation of linear rotational sequences of narrow resonances. The two-particle character of mesons or the three-particle character of baryons also falls within this category; if quark couplings were strong, hadrons would contain many $Q\bar{Q}$ pairs. As long as quark couplings are weak we can start from an approximation in which quarks are neglected. The criteria of the previous section would then apply.

Confinement as the word is usually understood depends on the existence of quantum numbers external to Q.C.D., such as ordinary electric charge, baryon number or flavor. One recognizes confinement as the absence of particles with certain values of these quantum numbers.

Thus, in a system with real quarks, the phase classification would be the same as in a system without quarks, provided we have conserved quantum numbers external to Q.C.D. However, the confinement phase would now have to be characterized by the absence of states with certain values of these quantum numbers.

The situation is altered somewhat if the quarks possess no external conserved quantum numbers. In particular, we may examine the Weinberg–Salam model (for SU(2), not SU(2)×U(1)), where the fundamental-representation particles, which we are calling quarks, acquire a vacuum-expectation value. The Weinberg–Salam phase, unlike a phase with Higgs breaking by adjoint-representation particles, *cannot support magnetic vortices*. We may therefore ask how we are to define it in a gauge-invariant

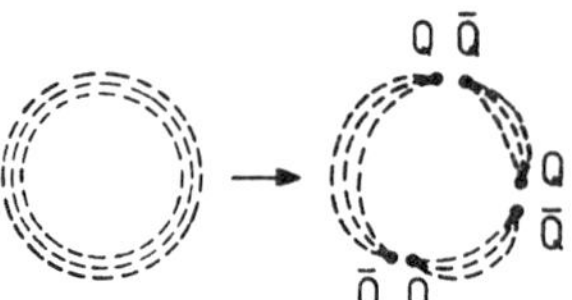

Fig. 3. Break-up of an electric vortex by the creation of $Q\bar{Q}$ pairs.

manner. If the coupling is weak, the meaning of the Weinberg–Salam phase is well understood. *But, in general, there appears to be no gauge-invariant way of defining the Weinberg–Salam phase, or of distinguishing it from the confinement phase.*

In certain models, the phase diagram contains regions with the usual distinguishing properties of each of the two phases, and one can pass continuously from one to the other without encountering a phase boundary. (Fradkin and Shenker [11], Banks and Rabinovici [12].)

Thus, for SU(2) systems with quarks (i.e., fundamental-representation particles), and no external conserved quantum numbers, the phase classification appears to be:

1) The perturbation-theory phase.
2) The Georgi–Glashow phase.
3) The phase with complete Higgs symmetry breaking by adjoint-representation particles (or, in general, by particles of any O(3) representation). This phase supports Nielsen–Olesen vortices; $\langle W\rangle = e^{-L}, \langle M\rangle = e^{-A}$.
4) The (Weinberg–Salam)-confinement phase; $\langle W\rangle = e^{-L}, \langle M\rangle = e^{-L}$.

We notice, in particular, that there is no precise distinction between confinement, with only colour singlets, and Higgs symmetry breaking, where the particles are not colour multiplets. This is because gauge invariance is not a symmetry in the usual sense.

5. The $1/N$ expansion

Due to space limitations this section will have to be curtailed; we refer to the article by Witten [13] for a fuller treatment. 't Hooft [14] has pointed out that Q.C.D. simplifies when

$$N \to \infty, \qquad g^2 N = \text{const.} \tag{15}$$

In this limit all hadrons are stable and quark couplings are weak. Since low-lying hadronic resonances are observed to be narrow, and useful models of hadrons can be constructed with neglect of $Q\bar{Q}$ pairs, it appears that the $1/N$ approximation is reasonably accurate for $N = 3$.

In two dimensions, 't Hooft [15] has solved Q.C.D. without making any approximation besides the $1/N$ approximation. $(\text{Q.C.D.})_2$ is thus a very useful guinea-pig for $(\text{Q.C.D.})_4$. Unfortunately confinement is automatic in two dimensions, since the Coulomb potential increases linearly with the distance. The two-dimensional model can therefore not help us understand confinement.

In four dimensions it has not yet been possible to obtain a solution without making approximations besides the smallness of $1/N$. Nevertheless, Q.C.D. in the $1/N$ limit is probably much simpler than the exact theory, and the $1/N$ expansion enables us to understand many features qualitatively.

6. Survey of confinement schemes

In the foregoing sections we have outlined the properties of the confinement phase and of other phases of gauge theories. We have not dealt with the problem of determining which phase of Q.C.D. actually corresponds to the physical vacuum, nor of calculating hadronic masses. Several schemes have been proposed for solving such problems. The schemes are not mutually exclusive, and the results are encouraging. Nevertheless, it is probably fair to say that, at the present time, no scheme has established confinement to the satisfaction of most workers in the field.

We shall mention five possible schemes, the list is not meant to be exhaustive.
1) Schemes based on instantons.
2) Schemes based on the magnetic instability of the vacuum.
3) Schemes based on equations for Green's functions.
4) Schemes based on lattice gauge theories.
5) Schemes based on strings.

1. *Schemes based on instantons*:

Instantons in three dimensions are long-range objects, i.e., the magnetic field falls off at least as slowly as r^{-2} at large distances. Polyakov [16] has shown that instantons can lead to confinement in a three-dimensional model.

In four dimensions instantons have a size parameter λ, and one must integrate over all values of λ. The integral is divergent at large values of λ and, at present, this divergence cannot be treated except by making a cut-off. If one does make such a cut-off, instantons are short-range objects; the magnetic field falls off like r^{-4}. Such short-range objects cannot give confinement.

The problem of the large instantons will probably be resolved, or shown to be irrelevant, at the same time as the confinement problem is solved. The question of which problem will help with the resolution of the other is at present open. Berg and Lüscher [17] have made some progress with the solution of the large-instanton problem.†

Though finite-size instantons cannot give actual confinement, they may come close. Callan, Dashen and Gross [18] (C.D.G.) suggest that there is an intermediate range of sizes where semi-perturbation methods, including instantons, may be applicable. The radii of the lighter hadrons may fall within this range. At the larger distances relevant for confinement perturbation methods would be inapplicable.

C.D.G. have constructed a theory of light hadrons, and have calculated the β-function, obtaining agreement with the lattice calculations as far as the strong-coupling region. Their results will form the subject of another talk, and I shall not describe them here.

For actual confinement, C.D.G. have proposed field distribution which they call "merons". These are long-range objects in four dimensions, three-dimensional cross-sections of which are monopoles. A four-dimensional meron plasma is probably equivalent to the monopole plasma mentioned earlier. A state of two merons has the topological properties of a single instanton, and C.D.G. conjecture that, if their instantons were allowed to split into two merons, they would obtain actual confinement.

Instantons may thus be a useful method of parametrizing the vacuum state, or of specifying the important configurations for functional integration. It should be mentioned, however, that the fundamental significance of instantons in a confined theory has been questioned [19].

2. *Methods based on the magnetic instability*:

Wilczek [20] and Saviddy [21] have noticed that one obtains an apparent lowering of the energy of a non-abelian system by giving the magnetic field a non-zero expectation value. The original calculation was performed in the one-loop approximation, but the result appears to be a fairly general feature of approximations in an infra-red unstable system [22]. The decrease in energy is of the form

$$\Delta E = -\text{const.}\ \mathscr{H}^2 \ln \mathscr{H}^2. \tag{16}$$

† Berg and Lüscher have solved the corresponding problem for the CP^{N-1} model. Due to lack of space we have not treated this model; there is no general agreement on the implications of the results on Q.C.D.

Owing to the presence of the logarithm, this contribution outweighs the classical contribution proportional to $\mathscr{H}^2$.

If Q.C.D. is the true theory of strong interactions the magnetic field cannot have a non-zero expectation value, since Lorentz-invariance would thereby be violated. Nevertheless, the apparent magnetic instability is probably indicative of some deep feature of Q.C.D.

N.K. Nielsen and Olesen [22] have found a further instability in the Q.C.D. vacuum with a non-zero magnetic field. H.B. Nielsen and Ninomiya [23] have shown that this instability can probably be removed by giving the vacuum a two-dimensional periodic structure, as in fig. 4. The magnetic field is perpendicular to the plane of the diagram, and is stronger inside than outside the circles. Each periodic region contains 2π unit of magnetic flux, and it may be regarded as a Nielsen–Olesen vortex. An alternative method of removing the instability, valid within a domain of finite size, has been suggested by Ambjorn, N.K. Nielsen and Olesen [24].

H.B. Nielsen and Olesen [25] have shown that quantum fluctuations will probably transform the crystal structure of Nielsen and Ninomiya into a liquid; equivalent results may be obtained from the domain structure suggested by the results of ref. [24]. The violation of Lorentz invariance is thus removed, and the flux tubes of Nielsen and Ninomiya become transformed into a "spaghetti". Their vacuum recalls a suggestion by 't Hooft [9]. One may obtain a superconductor, for which $\langle 0|\Phi|0\rangle \neq 0$, by constructing a coherent plasma of charged objects. Analogously, one might obtain a confined vacuum, for which $\langle 0|M|0\rangle$ is "large" (e^{-L}, not e^{-A}), by constructing a coherent plasma of Nielsen–Olesen vortices. In any case, the Wilson loop for such a vacuum is likely to behave like e^{-A} as a result of the commutation relations (13). The "Copenhagen vacuum" is thus very likely to confine.

3. *Methods based on equations for Green's functions*:

To link the approach now to be proposed with that of the previous subsection, we remark that the magnetic instability involves the *zero-frequency* component of $\mathscr{H}$, whereas confinement requires that the amplitude of the *low-frequency* components of $\mathscr{H}$ is larger than in the perturbation-theory vacuum. The two features therefore appear to be connected. Stated in another way, a non-zero expectation value of $\mathscr{H}$ corresponds to a ferromagnetic vacuum, whereas the confinement vacuum is a perfect para-magnet (Pagels and Tomboulis [26]). (Recall the superconductor-confinement analogy; in a superconductor, $\mu = 0$, in a confining vacuum, $\varepsilon = 0$. Since $\varepsilon\mu = 1$ in the vacuum, μ must be infinite.)

The connection between the magnetic instability and confinement may also be inferred from the fact that, if an applied magnetic field causes a lowering of the energy, a sufficiently spread-out monopole will be tachyonic. A monopole plasma, which confines, may therefore be formed.

Motivated by such considerations, we have attempted to approximate the equation for the gluon propagator in a manner which keeps as close as possible to the calculation of the energy of the vacuum in a static magnetic field [27]. The static field itself is replaced by the virtual gluon field. We obtain a

Fig. 4. Cross section of the Nielsen–Ninomiya vacuum.

simple truncation of the Schwinger–Dyson equations. Rough numerical solution yields a gluon propagator which behaves like p^{-4} at low momenta.

A slightly different truncation scheme for the Schwinger–Dyson equations has been proposed by Anishetty, Baker, Ball, Kim and Zachariasen [28].

Naïve power counting indicates that the exchange of a p^{-4} propagator, would give an extra factor x^2 in the $Q\bar{Q}$ potential. In other words, the Coulomb potential would be replaced by a potential proportional to $|x|$, and we obtain confinement. To put this argument on a more quantitative footing and to perform calculations, we have made two further approximations. A behaviour p^{-4}, which we have *obtained* from our equations, is that which would be expected from a Wilson loop which behaves like e^{-A}. We *assume* similar results for multi-gluon Green's functions. Our assumption should be checked for the four-point Green's function.

We are then led to a Bethe–Salpeter equation with fictitious "abelian gluons". Diagrams with *crossed fictious* gluons correspond to diagrams with *real*, interacting, *uncrossed* non-abelian gluons. (Diagrams with real crossed gluons do not appear in the lowest order of the $1/N$ approximation.) The solution of a Bethe–Salpeter equation with crossed and uncrossed diagrams is itself a formidable problem. It may be simplified if we make a non-covariant approximation for the Wilson loop; we divide the loop into strips by lines perpendicular to the time axis, and project each strip into a plane containing the time axis. We might estimate the error by considering a pair of quarks at the ends of a rotating string; our estimate is then incorrect by a factor $\pi/4$. We have not examined the corrections to the approximation, but they will probably involve vibrational modes of a "string".

Our approximations lead to a static potential between two quarks proportional to the distance between them. Such a force gives rise to confinement and to linearly rising Regge trajectories.

The possibility of chiral symmetry breaking is inherent in our equations; whether such symmetry breaking actually occurs cannot be decided without explicit numerical solution. A zero-mass pion will automatically occur if there is chiral symmetry breaking. In a calculation with closed fermion loops, the flavour-singlet sector will contain no zero-mass pseudoscalar particle, even if instantons are not explicitly included. This feature is in agreement with expectations by Witten [19] and others.

4. *Direct calculations with lattice-gauge theories* have now been taken to a point where the fundamental lattice size is small enough to fall within the perturbation region, and where the theory is therefore a good approximation to continuum theory. Since the very impressive results so far obtained are a major subject of this meeting I shall not include them in my talk.

5. *String formulations*:

Several groups are presently attempting to reformulate gauge theories directly in terms of the "string" operator $:\exp\{ig \int dx^\mu \tau^a A^a_\mu(x)\}:$ joining a $Q\bar{Q}$ pair, or of the corresponding Wilson-loop operator for pure gauge theories. The formulation has the advantage of never referring to unphysical non-gauge-invariant quantities. As far as I am aware, the only non-perturbative calculations to date are those by Gervais and Neveu. As the authors themselves will present their results I shall omit these, too, from my talk.

References

[1] Dual Theory, ed. M. Jacob (North-Holland, Amsterdam, 1974).
[2] H.B. Nielsen and P. Olesen, Nucl. Phys. B61 (1973) 45.
[3] Y. Nambu, Phys. Rev. D10 (1974) 4262.

[4] S. Mandelstam, Phys. Reports 23C (1976) 245.
[5] G. 't Hooft, in: High Energy Physics, Proc. European Phys. Soc. Int. Conf., ed. A. Zichichi (Editrice Compositori, Bologna, 1976) p. 1225.
[6] T.T. Wu and C.N. Yang, in: Properties of Matter under Unusual Conditions, eds. H. Mark and S. Fernbach (Interscience, New York, 1969) p. 349.
[7] G. 't Hooft, Nucl. Phys. B79 (1976) 276.
[8] K.G. Wilson, Phys. Rev. D10 (1974) 2445.
[9] G. 't Hooft, Nucl. Phys. B138 (1978) 1.
[10] S. Mandelstam, Phys. Rev. D19 (1979) 2391.
[11] E. Fradkin and S. Shenker, Phys. Rev. D19 (1979) 3682.
[12] T. Banks and E. Rabinovici, Nucl. Phys. B160 (1979) 349.
[13] E. Witten, Nucl. Phys. B160 (1979) 57.
[14] G. 't Hooft, Nucl. Phys. B72 (1974) 461.
[15] G. 't Hooft, Nucl. Phys. B75 (1974) 461.
[16] A.M. Polyakov, Nucl. Phys. B120 (1977) 429.
[17] B. Berg and M. Lüscher, Nucl. Phys. B160 (1979) 281.
[18] C.G. Callan, R.F. Dashen and D.J. Gross, Phys. Rev. D17 (1978) 2717; D19 (1979) 1826; D20 (1979) 3279.
[19] E. Witten, Nucl. Phys. B149 (1979) 285.
[20] F. Wilczek, private communication from D. Gross.
[21] G.K. Saviddy, Phys. Lett. 71B (1977) 133.
[22] N.K. Nielsen and P. Olesen, Nucl. Phys. B144 (1978) 376.
[23] H.B. Nielsen and M. Ninomiya, Nucl. Phys. B156 (1979) 1.
[24] J. Ambjørn, N.K. Nielsen and P. Olesen, Nucl. Phys. B152 (1979) 75.
[25] H.B. Nielsen and P. Olesen, Nucl. Phys. B160 (1979) 380.
[26] H. Pagels and E. Tomboulis, Nucl. Phys. B143 (1978) 485.
[27] S. Mandelstam, Phys. Rev. 20 (1979) 3223.
[28] R. Anishetty, M. Baker, J.S. Ball, S.K. Kim and F. Zachariasen, Phys. Lett. 86B (1979) 52;
M. Baker, private communication.

VOLUME 31, NUMBER 10 PHYSICAL REVIEW LETTERS 3 SEPTEMBER 1973

Is Baryon Number Conserved?

Jogesh C. Pati*
International Centre for Theoretical Physics, Trieste, Italy, and Department of Physics and Astronomy, University of Maryland, College Park, Maryland 20742

and

Abdus Salam
International Centre for Theoretical Physics, Trieste, Italy, and Imperial College, London, England
(Received 3 August 1973)

We suggest that baryon-number conservation may not be absolute and that an integrally charged quark may disintegrate into two leptons and an antilepton with a coupling strength $G_B m_p^2 \lesssim 10^{-9}$. On the other hand, if quarks are much heavier than low-lying hadrons, the decay of a three-quark system like the proton is highly forbidden (proton lifetime $\gtrsim 10^{28}$ y). Motivation for these ideas appears to arise within a unified theory of hadrons and leptons and their gauge interactions. We emphasize the consequences of such a possibility for real quark searches.

It is part of general belief in particle physics that conservation of baryon number is an absolute law of nature. Such a notion is but natural when one considers the extraordinary stability of the lightest known baryon, the proton, with a lifetime in excess of 10^{35} sec. In this note we wish to question whether this apparent proton stability truly reflects the conservation of baryon number to a similar degree.

Specifically, we have in mind the following possibility. Assume that the proton is made up in some sense of three quarks, each quark (q) carrying baryon number $B = 1$ and an integral electric charge. Assume that quarks and diquarks (if the latter exist) are heavier than the low-lying hadrons. Assume further that a quark can decay into two of the known leptons ($l = \nu_e, e^-, \nu_\mu, \mu^-$) and an antilepton, the decay being described by an effective Lagrangian

$$\mathcal{L}_{\text{eff}} = (G_B/\sqrt{2})(\bar{q}l)(\bar{l}l) + \text{H.c.} \tag{1}$$

This decay violates conservation of baryon and lepton (L) numbers, but conserves fermion number F where $F = B + L$.

Our point is this: If states with quark and diquark quantum numbers are heavier than the low-lying hadrons, a proton ($B = 3$, $F = 3$) can have a real decay only to three leptons plus mesons, or four leptons plus an antilepton, etc. Since this involves a violation of baryon number by three units, the lowest-order amplitude in G_B in which a proton can decay[1] is therefore G_B^3. Assume that $G_B m_p^2 \lesssim 10^{-9}$. (We give later our theoretical reasons for G_B being less than the decay constant for $|\Delta S| \neq 0$ neutral semileptonic transitions, which is of order $G_F\alpha^2$ empirically.) We then find that the decay $q \to l + l + \bar{l}$ may be associated with a lifetime as short as 10^{-10} sec, say (depending on quark mass), whereas the proton's lifetime could still be far in excess of 10^{35} sec because of the high degree of forbiddenness of its decay. Such a model would therefore show that (i) quarks (if they exist) may exhibit unexpected decay properties involving violation of baryon number as well as lepton number without conflicting with the observed degree of stability of the proton, and (ii) there is the possibility that there is no stable quark, contrary to present belief. This may be one reason why conventional searches for quarks have been unsuccessful (especially if $\tau_q \lesssim 10^{-10}$ sec). (Note that if quarks were fractionally charged, electric-charge con-

VOLUME 31, NUMBER 10 PHYSICAL REVIEW LETTERS 3 SEPTEMBER 1973

servation would imply that a stable quark must exist unless there were lighter fractionally charged leptons[2] into which the quarks could decay.)

Before giving the motivation which led us to consider an effective interaction of the type (1) —*and we stress that the general considerations above hold irrespective of any specific model* —we give some order-of-magnitude estimates for typical quark and proton decay widths (see Table I). The quark decay width follows directly from Eq. (1):

$$\Gamma(q \to l+l+\bar{l}) = G_B{}^2 m_q{}^5/24(2\pi)^3 = \lambda^2(3\times10^7\ \mathrm{sec}^{-1})\left(\frac{m_q}{10\ \mathrm{GeV}}\right)^5, \quad (2)$$

where we have put $G_B = \lambda G_F \alpha^2$ (we estimate $\lambda \lesssim 1$; see below). Two typical proton decay amplitudes (M) and corresponding rates ($\Gamma = |M|^2\rho$, where ρ is the phase-space factor) are listed in Table I. The quantities m_q and m_p are quark[3] and proton masses; A_4 and A_5 are numerical factors arising out of four- and five-particle phase-space integrations, which depend somewhat on precise matrix elements [usually they are $\gg 1$; for example $A_3 = 12$, see Eq. (2)]; Λ is a cutoff, which may be of the order of a few GeV (as in the K_L-K_S mass-difference calculation). Using Table I, we obtain for a typical proton decay[4]

$$\Gamma(p \to 3l + \pi) = 2\times10^{-39}\lambda^6/A_4\ \mathrm{sec}^{-1} \quad (3)$$

for $m_q = 10$ GeV and $\Lambda = 2m_p$. Thus for $\lambda \lesssim 1$, $\Gamma_p \lesssim 2\times10^{-39}\ \mathrm{sec}^{-1}$. Although one may not take the precise estimates given above too seriously, the main point we wish to emphasize is that the high degree of forbiddenness of the proton decay relative to the quark decay appears to be sufficient to lay open the possibility that the proton may be comfortably stable ($\tau_p > 10^{28}$ yr) and yet the quarks decaying into leptons sufficiently short lived.

Since $G_B m_p{}^2 \approx 10^{-9}$ implies a characteristic energy (or intermediate meson mass) of the order of 3×10^4 GeV, one may expect that at (cosmic-ray) energies of this order, reaction rates for the processes $e+p$ or $p+p \to$ leptons + antileptons, etc., would attain unitarity limit and effectively become strong. Thus a study of multilepton-induced showers at high cosmic-ray energies may be one way of testing the ideas presented above. A clearer test could hopefully be provided by extending the search for real integrally charged quarks at the CERN intersecting storage rings and the National Accelerator Laboratory to detect possible disintegration of quarks into *energetic* leptons ($q \to l+l+\bar{l}$, $q\to l+\pi$, $q \to l+\gamma$, ..., etc.). One should allow for $\tau_q \gtrsim 10^{-10}$ sec on the one hand and perhaps[5] as short as 10^{-12} to 10^{-13} sec on the other.

Our basic motivation for B nonconservation comes from a recent attempt[1,6] at a gauge theory of strong, weak, and electromagnetic interactions. To construct a unified, anomaly-free, renormalizable gauge model we suggested that a system of twelve integrally charged quarks (nine of Han-Nambu variety, and three charmed quarks) plus the four known leptons ($\nu_e, e^-, \mu^-, \nu_\mu$) be combined in a $(\underline{4}, \underline{4}^*)$ representation F of an $\mathrm{SU}(4)_{L+R}{}' \otimes \mathrm{SU}(4)_{L+R}{}''$ group structure,

$$F = \begin{pmatrix} \mathcal{P}_a{}^0 & \mathcal{P}_b{}^+ & \mathcal{P}_c{}^+ & \nu_e \\ \mathfrak{N}_a{}^- & \mathfrak{N}_b{}^0 & \mathfrak{N}_c{}^0 & e^- \\ \Lambda_a{}^- & \Lambda_b{}^0 & \Lambda_c{}^0 & \mu^- \\ \chi_a{}^0 & \chi_b{}^+ & \chi_c{}^+ & \nu_\mu \end{pmatrix}. \quad (4)$$

The strong interactions were introduced by gauging an $\mathrm{SU}(3)_{L+R}{}''$ subgroup of $\mathrm{SU}(4)_{L+R}{}''$, the conventional weak interactions by gauging an $[\mathrm{SU}_L(2)']$ subgroup of $\mathrm{SU}(4)_L{}'$, while the electromagnetic gauges spanned over generators of *both* SU(4)′ and SU(4)″, i.e.,

$$Q = (I_3{}' + \tfrac{1}{2}Y' - \tfrac{2}{3}C')_{L+R} + (I_3{}'' + \tfrac{1}{2}Y'' - \tfrac{2}{3}C'')_{L+R}. \quad (5)$$

TABLE I. Estimates of typical proton decay modes. We have not exhibited factors of $(2\pi)^{-n}$ ($n>0$) in the matrix element M, which usually arise from virtual loops. These suppress proton decay rate still further.

Decay	M	ρ	Γ
$p \to l+l+l+\pi$	$\left(\frac{G_B}{\sqrt{2}}\right)^3\left(\frac{\Lambda}{m_q}\right)^3\Lambda^3$	$\frac{M_p{}^7}{A_4(2\pi)^5}$	$m_p\left(\frac{G_B m_p{}^2}{\sqrt{2}}\right)^6\left(\frac{\Lambda}{m_q}\right)^6\left(\frac{\Lambda}{m_p}\right)^6\frac{1}{A_4(2\pi)^5}$
$p \to l+l+l+l+\bar{l}$	$\left(\frac{G_B}{\sqrt{2}}\right)^3\left(\frac{\Lambda}{m_q}\right)^3\Lambda$	$\frac{M_p{}^{11}}{A_5(2\pi)^7}$	$m_p\left(\frac{G_B m_p{}^2}{\sqrt{2}}\right)^6\left(\frac{\Lambda}{m_q}\right)^6\left(\frac{\Lambda}{m_p}\right)\frac{1}{A_5(2\pi)^7}$

VOLUME 31, NUMBER 10 PHYSICAL REVIEW LETTERS 3 SEPTEMBER 1973

The theory at this stage had no exotic consequences except for the unusual unification of hadronic matter ($B=1$, $L=0$) with leptonic matter ($B=0$, $L=1$) within the *same* multiplet of a common symmetry structure $SU(4)'\otimes SU(4)''$.

But in order that such a unification be dynamically compelling, one must gauge sufficient degrees of freedom (consistent with established[7] selection rules) to ensure transformability of leptons into baryons. This still does not imply nonconservation of baryon-lepton numbers because appropriate gauge bosons could carry these numbers. What we want to show is that if in addition to the subgroups mentioned above we had also gauged the remaining degrees of freedom of $SU(4)'$ and $SU(4)''$, or even a non-Abelian subset (stated below) such that the electric current is expressed as a sum of *non-Abelian* currents from both groups $SU(4)'$ and $SU(4)''$, the requirement of electric-charge conservation—expressed in terms of masslessness of the photon—together with the twin requirements of renormalizability and appropriate[8] massiveness of all other gauge bosons, necessarily appears to lead to lepton-baryon-number nonconservation.

To demonstrate this, consider the local gauge structure[9,10] $SU(4)_L'\otimes SU(4)_R'\otimes SU(4)_{L+R}''$ {although the essential ingredients of the argument become manifest already at the stage of the smaller gauge symmetry $[SU(2)_L']\otimes[SU(2)_R']\otimes SU(4)_{L+R}''$, preferable for reasons connected with anomalies}. Let $W_{ij}{}^{L,R}$ and V_{ij} ($i,j=1, 2, 3, 4$) represent the fifteen-plet of gauge mesons associated with the groups $SU(4)_{L,R}'$ and $SU(4)_{L+R}''$ with $J_{ij}{}'^{L,R}$ and J_{ij}'' denoting the associated currents. Note that the quantum numbers B, L associated with these currents are as follows:

$J_{ij}{}'^{L,R}$ (all i and j), $B=L=0$;

J_{ij}'' ($i,j=1, 2, 3$; $i=j=4$), $B=L=0$;

J_{ij}'' ($i=1, 2, 3$; $j=4$), $B=1$, $L=-1$;

J_{ij}'' ($i=4$; $j=1, 2, 3$), $B=-1$, $L=+1$;

the last two groups being exotic. The point to be emphasized is that unless the theory forces a mixing of the nonexotic currents ($B=L=0$) with the exotic ones (with $B\neq 0$, $L\neq 0$), the mere existence of such currents and the corresponding gauge mesons would not violate B-L conservation. Such a mixing, however, appears necessary if one attempts to give masses to all gauge mesons (with the sole exception of the photon) through a Higgs-Kibble mechanism. This is because with the electric charge expressed as a sum of non-Abelian generators [as given by Eq. (5)], appropriate massiveness of all gauge bosons other than the photon can be secured only[11] by postulating, among other representations, the existence of a mixed representation of Higgs-Kibble σ particles—typically a $(\underline{1}, \underline{4}, \underline{4}^*)$ representation of $SU(4)_L'\otimes SU(4)_R'\otimes SU(4)_{L+R}''$—with expectation values in the sequence indicated:

$$\langle\sigma\rangle_{ij}=\alpha_i\delta_{ij}\quad (i,j=1, 2, 3, 4). \tag{6}$$

Quite clearly, the gauge term in the Lagrangian $|g_R W^R\langle\sigma\rangle+f\langle\sigma\rangle V|^2$ [where g_R and f are the coupling constants associated with weak $SU(4)_R'$ and strong $SU(4)_{L+R}'$ gauge groups, respectively] induces not only the appropriate mixing of neutral V's and the W's which go to make up the photon but also a mixing of the exotic V's with W^R's, coupled to $B=0=L$ currents. It is this mixing which is responsible for B-L nonconservation.

A typical term in the effective current-current Lagrangian induced by the above mixing is of the form

$$\mathcal{L}_B=G_B(\bar{\mathcal{P}}_a{}^0\nu_e+\bar{\mathfrak{N}}_a{}^-e^-+\bar{\lambda}_a{}^-\mu^-+\bar{\chi}^0\nu_\mu)\times(C^0+\bar{\nu}_\mu\nu_e)_R+\text{H.c.}, \tag{7}$$

where $C^0=(\bar{\mathcal{P}}_a{}^0\chi_a{}^0+\bar{\mathcal{P}}_b\chi_b{}^++\bar{\mathcal{P}}_c\chi_c{}^+)_R$ is the charm current and $\mathcal{L}_B$ is part of the structure

$$\alpha_1\alpha_4J_{14}''J_{41}'^R+\alpha_2\alpha_4J_{24}''J_{42}'^R+\alpha_3\alpha_4J_{34}''J_{43}'^R+\text{H.c.} \tag{8}$$

To obtain an estimate of G_B, first note that exchanges of the exotic V mesons [as well as exchanges of $W_{3i}{}^R$ ($i=1, 2$)] induce neutral semileptonic $|\Delta S|\neq 0$ transitions with effective strength $\simeq f^2/m_x{}^2$, where m_x is an exotic meson mass. In order that this be consistent with observed limits, $f^2/m_x{}^2$ must be $\lesssim G_F\alpha^2$. Thus[12] $m_x\gtrsim f(3\times 10^4$ GeV). Since $G_B=\kappa(f^2/m_x{}^2)$, where κ is a mixing parameter (whose detailed value depends on the mass matrix), we infer that empirically $G_B\lesssim G_F\alpha^2$.

It is amusing to note that, depending on the details of the model chosen for the Higgs-Kibble scalars (and whether μ^- or e^- is the "strange" lepton), one will encounter varying selection rules for quark and proton decays. The precise structure of the B-L-nonconserving interaction obtained above leads to quark decays of the fol-

VOLUME 31, NUMBER 10 PHYSICAL REVIEW LETTERS 3 SEPTEMBER 1973

lowing variety:

(A) $(\mathcal{P}_a^0, \mathfrak{N}_a^-, \lambda_a^-) \to (\nu_e, e^-, \mu^-) + (\nu_\mu + \bar{\nu}_e)$.

(B) $(\mathcal{P}_b^+, \mathfrak{N}_b^0, \lambda_b^0) \to (\nu_e, e^-, \mu^-) + (\nu_\mu + e^+)$.

(C) $(\mathcal{P}_c^+, \mathfrak{N}_c^0, \lambda_c^0) \to (\nu_e, e^-, \mu^-) + (\nu_\mu + \mu^+)$.

These would lead to a proton's decay to seven or nine leptons (including antileptons), but not to $3l + \pi$ or $4l + \bar{l}$. One may note that within the smaller gauge structure $\{[SU(2)_L'] \otimes [SU(2)_R'] \otimes SU(4)_{L+R}''\}$, B-L nonconservation proceeds only through the term $J_{34}''(J_{43}'^{L} + J_{12}'^{R})$. This will allow decays of the type (C), but not of (A) and (B). In this case, since the proton is made up of (a, b, c) quarks, one can show that its decay is further suppressed[13] at least by additional factors of α.

Since the characteristic energies ($\approx 10^4$–10^5 GeV) discussed above (which, we stress, represent a *new scale* in particle physics) are not the energies encountered in normal star interiors, one does not expect significant astrophysical implications of B-L nonconservation except in the early stages of the universe (when baryons may have been produced from energetic lepton-lepton collisions or vice versa) and possibly in black-holes and quasars.

To conclude, while arguments based on a particular set of theoretical ideas are never compelling, the general considerations on forbiddenness of proton decay in a heavy quark model remain and need experimental verification. If the gauge ideas are correct, we find it amusing that the only known massless gauge particle is the photon. Could it be that the electric charge is the only non-Abelian[14] conserved charge in nature?

We thank Professor P. Budini, Professor R. Dalitz, Professor R. Glasser, Professor M. Goldhaber, Professor D. Sciama, Professor A. Sirlin, and Dr. J. Strathdee for discussions. J.C.P. wishes to acknowledge the hospitality of the International Atomic Energy Agency and UNESCO during his stay at the International Centre for Theoretical Physics, Trieste, in the summer of 1973.

---

*Work supported in part by the National Science Foundation under Grant No. NSF GP 20709.

[1]This remark orginates from Appendix A of J. C. Pati and A. Salam, Phys. Rev. D 8, 1240 (1973). The motivation for baryon-lepton–number nonconservation in the present paper is very different from that presented in this Appendix.

[2]A. Salam and J. C. Pati, Phys. Lett. 43B, 311 (1973).

[3]The characteristic appearance of $(G_B/m_q)^3$ in the proton-decay matrix elements is due to the fact that the unitarity sum for ImM necessarily contains a product of the three physical-quark–decay matrix elements and hence the real part is proportional to $\sum_E G_B{}^3(m_q - E)^{-3}$.

[4]Note that of the value given in Eq. (3), 6×10^{-35} sec^{-1} comes from a characteristic three-body decay width alone (if such a decay were allowed) i.e., $\Gamma(p \to 3l) = (96\pi^3)^{-1}(2^{-3/2}G_B{}^3 m_p{}^6)^2 m_p$, with $G_B m_p{}^2 = G_F \alpha^2$.

[5]Because of the uncertainty in the estimate of the proton-decay matrix element, it is possible that the quark lifetime could even be of order 10^{-12} to 10^{-13} sec without conflicting with the proton lifetime, especially if additional selection rules are involved in proton decay (see also remarks below).

[6]C. Itoh, T. Miamikawa, K. Miura, and T. Watanabe, "Unified Gauge Theory of Weak Electromagnetic and Strong Interactions" (to be published). This model is similar to that of Ref. 1, except that quarks are fractionally charged while leptons are integrally charged and the gauge bosons are massless. There would be no possibility of quarks decaying into leptons in this scheme.

[7]The whole purpose of the introductory section was to question whether baryon conservation is indeed all that well established.

[8]Consistent with approximate global SU(3)'' symmetry and effectively weak lepton interactions [so that $\alpha_1 \simeq \alpha_2 \simeq \alpha_3 \neq 0$ and $\alpha_4 \neq 0$ and large in Eq. (6)].

[9]The desirability of gauging an extended group structure was suggested in Ref. 1. The bigger group structure $SU(4)_L' \otimes SU(4)_R' \otimes SU(4)_{L+R}''$ leads to anomalies. On the other hand a simple and elegant scheme is obtained within the smaller gauge symmetry $SU(2)_L' \otimes SU(2)_R' \otimes SU(4)_{L+R}''$, which we consider in some detail in a forthcoming note.

[10]D. Ross, to be published, has independently considered the consequences of gauging $SU(4)' \times SU(4)''$ within the unified model of Ref. 1. His work confirms the conclusion regarding B-L nonconservation in such a scheme.

[11]The necessity for such a representation involves a longer discussion and will be given elsewhere.

[12]It is well known that suppression due to large masses is not retained in general by loop diagrams (see for example, S. Weinberg, Phys. Rev., to be published. Preliminary studies reveal that in the model presented here B and L nonconservations are suppressed by heavy masses even in loop diagrams. This is to be considered in detail elsewhere.

[13]It is even possible that the proton could be made absolutely stable in this model, provided there exists an additional particle (meson) in the theory, which is also absolutely stable and heavier than the proton (see Appendix A of Ref. 1 for details of this mechanism).

[14]There is, of course, still the possibility of gauging the U(1) Abelian generator, corresponding to fermion-number conservation in the theory.

VOLUME 32, NUMBER 8 PHYSICAL REVIEW LETTERS 25 FEBRUARY 1974

Unity of All Elementary-Particle Forces

Howard Georgi* and S. L. Glashow
Lyman Laboratory of Physics, Harvard University, Cambridge, Massachusetts 02138
(Received 10 January 1974)

Strong, electromagnetic, and weak forces are conjectured to arise from a single fundamental interaction based on the gauge group SU(5).

We present a series of hypotheses and speculations leading inescapably to the conclusion that SU(5) is the gauge group of the world—that all elementary particle forces (strong, weak, and electromagnetic) are different manifestations of the same fundamental interaction involving a single coupling strength, the fine-structure constant. Our hypotheses may be wrong and our speculations idle, but the uniqueness and simplicity of our scheme are reasons enough that it be taken seriously.

Our starting point is the assumption that *weak and electromagnetic forces are mediated by the vector bosons of a gauge-invariant theory with spontaneous symmetry breaking.* A model describing the interactions of leptons using the gauge group $SU(2)\otimes U(1)$ was first proposed by Glashow, and was improved by Weinberg and Salam who incorporated spontaneous symmetry breaking.[1] This scheme can also describe hadrons, and is just one example of an infinite class of models compatible with observed weak-interaction phenomenology. If we assume that *there are as few fermion fields as possible* and, in particular, that there are no unobserved leptons, the Weinberg model becomes unique up to extensions of the gauge group: The observed leptons may be described by six left-handed Weyl fields $(e_L^-, \mu_L^-, \nu_L, \nu_L', e_L^+, \mu_L^+)$ and their charge conjugates. If the gauge couplings do not mix leptons with quarks, these six fields must transform as a representation of the gauge group: one of the 23 subgroups of U(6) containing an $SU(2)\otimes U(1)$ subgroup in which the leptons behave as they do in the Weinberg model.

To include hadrons in the theory, we must use the Glashow-Iliopoulos-Maiani (GIM) mechanism and introduce a fourth quark p' carrying charm.[2] Still, decisions must be made: Should the quarks have fractional or integer charges? Should there be one quartet of quarks or several? Bouchiat, Iliopoulos, and Meyer suggested what seems the most attractive alternative: *three quartets of fractionally charged quarks.*[3] This combination of the GIM mechanism with the notion of colored quarks[4] keeps the successes of the quark model and gives an important bonus: Lepton and hadron anomalies cancel so that the theory of weak and electromagnetic interactions is renormalizable.[5]

The next step is to include strong interactions. We assume that *strong interactions are mediated by an octet of neutral vector gauge gluons* associated with local color SU(3) symmetry, and that there are no fundamental strongly interacting scalar-meson fields.[6] This insures that parity and hypercharge are conserved to order α,[7] and does not lead to any new anomalies, so that the theory remains renormalizable. The strongest binding forces are in color singlet states which may explain why observed hadrons lie in qqq and $q\bar{q}$ configurations.[8] And, it gives another important bonus: Since the strong interactions are associated with a non-Abelian theory, they may be asymptotically free.[9]

Thus, we see how attractive it is for strong, weak, and electromagnetic interactions to spring from a gauge theory based on the group $\mathfrak{F} = SU(3)\otimes SU(2)\otimes U(1)$. Alas, this theory is defective in one important respect: It does not truly unify weak and electromagnetic interactions. The $SU(2)\otimes U(1)$ gauge couplings describe two interactions with two independent coupling constants; a true unification would involve only one.

Electric charge is observed to be quantized. This has no natural explanation in the framework of conventional quantum electrodynamics, but it is necessarily true in any unified theory[10]—yet another reason to search for a true unification.

We must assume that the gauge group is larger than $\mathfrak{F}$. Suppose it is of the form $SU(3)\otimes\mathfrak{W}$ where $\mathfrak{W}$ contains $SU(2)\otimes U(1)$ but has a unique gauge coupling constant. $\mathfrak{W}$ must be simple, or the direct product of isomorphic simple factors with discrete symmetries which interchange them. This embedding of the Weinberg model implies a relationship between the coupling constants of the SU(2) and U(1) subgroups. Because leptons are singlets under color SU(3), leptons and quarks

Volume 32, Number 8 PHYSICAL REVIEW LETTERS 25 February 1974

must lie in separate representations of $\mathcal{W}$. If only the six observed lepton states are involved, $\mathcal{W}$ must be one of the 23 relevant subgroups of U(6). The only candidates involving a single gauge coupling constant are SU(3), SU(3)⊗SU(3), and SU(6).[11] For each of these cases, the mixing angle is fixed so that $\sin^2\theta_w = \frac{1}{4}$. None of these schemes can describe hadrons: The generator corresponding to electric charge does not admit fractional charges, nor, being traceless, can it explain why the sum of the quark charges is not zero. No gauge group of the form SU(3) ⊗$\mathcal{W}$ works.

We see that we cannot unify weak and electromagnetic interactions independently of strong interactions. The remaining possibility is that the gauge group $\mathcal{G}$ contains $\mathcal{F}$ as a subgroup but is itself simple or the direct product of isomorphic simple factors. Leptons and quarks must lie together in the same irreducible representations of such a group: Some gauge fields carry lepton number and quark number. The same coupling strength—the fine-structure constant—characterizes all three kinds of interaction. This outrageous possibility may seem palatable after the following discussion about asymptotic freedom and its complement, infrared slavery.

Asymptotic freedom is a property of non-Abelian Yang-Mills field theories which promises to explain the pointlike structure of hadrons at high energy.[9] Unfortunately, these theories do not appear to describe strong interactions correctly since they involve massless strongly interacting vector bosons. The obvious solution is to introduce strongly interacting scalar mesons which develop vacuum expectation values, spontaneously break the gauge symmetry, and generate vector meson masses by the Higgs mechanism. But the scalar-meson Lagrangian involves additional renormalizable couplings which may spoil asymptotic freedom. Sadly, no one has found an asymptotically free model in which the gauge symmetry is completely broken and all the vector mesons develop mass.[12]

Weinberg, and Gross and Wilczek,[13] propose an astonishingly radical solution: to leave the gauge symmetry unbroken. *While the Yang-Mills Lagrangian appears to describe massless vector bosons, the hideous infrared divergences of the theory conspire to prevent their appearance in physical states.* This could explain the absence of physical states that are not color singlets and answer the old saw: Why don't the quarks get out? We have nothing to say about the merits of this picture; we assume that it works and use it.

The essential thing about a theory of strong interactions based on an unbroken non-Abelian gauge symmetry is that the strength of strong interactions no longer depends on the existence of a large coupling constant. Even if the gauge coupling constant is small, say of order e, the infrared divergences of the theory can lead to phenomenological interactions strong enough to keep the quarks bound.[14] *What we want is not asymptotic freedom but infrared slavery.*

The theory we have in mind involves a unifying gauge group $\mathcal{G}$ whose only coupling constant is the unit of electric charge and which contains—in an appropriate way—the subgroup $\mathcal{F}$. The symmetry is spontaneously broken leaving only the direct product of color SU(3) and electromagnetic gauge invariance as exact local symmetries. Color SU(3) is an unbroken non-Abelian gauge symmetry causing infrared slavery and leading to strong interactions. Electromagnetic gauge invariance is Abelian and commutes with color SU(3). Since the photon has no direct couplings to the gauge fields of color SU(3), electromagnetism is free of insolvable infrared-divergence problems, and photons may be freely emitted and absorbed. All other gauge fields develop masses through the Higgs mechanism. Those associated with the subgroup SU(3)⊗U(1), aside from the photon, mediate ordinary weak interactions and the neutral-current effects of the Weinberg model. The rest, which are colored and massive, mediate new and presumably even weaker interactions.

Our unifying group $\mathcal{G}$ must be of rank at least 4. There are exactly nine rank-4 local Lie groups which can involve only one coupling strength: $[SU(2)]^4$, $[O(5)]^2$, $[SU(3)]^2$, $[G_2]^2$, O(8), O(9), Sp(8), F_4, and SU(5).' The first two are unacceptable since they do not contain SU(3). To proceed, we review the behavior of quarks and leptons under $\mathcal{F}$.

We use the Weyl notation in which all fermion fields are left-handed two-component spinors. There are thirty such fields in our picture of nature: four leptons $(\mu^-, \nu', e^-, \nu)_L$, two antileptons $(\mu^+, e^+)_L$, twelve quarks $(p_i', p_i, n_i, \lambda_i)_L$, and twelve antiquarks $(\bar{p}_i', \bar{p}_i, \bar{n}_i, \bar{\lambda}_i)_L$, where the color index i assumes three values. Under the subgroup SU(3)⊗SU(2), the leptons are SU(3) singlets and SU(2) doublets; the antileptons are singlets under both groups; the quarks are SU(3) triplets as well as SU(2) doublets; and, finally, the anti-quarks are SU(3) $\underline{3}$*'s but SU(2) singlets.

VOLUME 32, NUMBER 8 PHYSICAL REVIEW LETTERS 25 FEBRUARY 1974

The $SU(3)\otimes SU(2)$ content of the thirty fields is $2(\underline{1},2)\oplus 2(\underline{1},1)\oplus 2(\underline{3},2)\oplus 4(\underline{3}^*,1)$ in an evident notation.

This representation is complex, not equivalent to its complex conjugate. So also is the corresponding representation of $\mathcal{G}$. Of our nine candidates only $[SU(3)]^2$ and SU(5) admit complex representations. We have already considered and rejected $[SU(3)]^2$ in our discussion of the synthesis of just weak and electromagnetic interactions. We are left with SU(5).

Under the subgroup $SU(3)\otimes SU(2)$, the fundamental five-dimensional representation of SU(5) transforms like $(\underline{1},2)\oplus(\underline{3},1)$. The complex conjugate $\underline{5}^*$ transforms like $(\underline{1},2)\oplus(\underline{3}^*,1)$. The irreducible ten-dimensional representation given by the antisymmetrized tensor product of two $\underline{5}$'s transforms like $(\underline{1},1)\oplus(\underline{3}^*,1)\oplus(\underline{3},2)$. If the thirty left-handed fermions transform like two $\underline{10}$'s and two $\underline{5}^*$'s, the $\mathcal{F}$ content is just right to describe physics. In order to display these representations, we replace the two $\underline{5}^*$'s of left-handed fields by two $\underline{5}^*$'s of their right-handed charge conjugates. The representations containing electrons are then a $\underline{5}$ and a $\underline{10}$:

$$\begin{bmatrix} n_1 \\ n_2 \\ n_3 \\ e^+ \\ \nu \end{bmatrix}_R, \quad \frac{1}{\sqrt{2}}\begin{bmatrix} 0 & \bar{p}_3 & -\bar{p}_2 & -p_1(\theta) & -n_1 \\ -\bar{p}_3 & 0 & \bar{p}_1 & -p_2(\theta) & -n_2 \\ \bar{p}_2 & -\bar{p}_1 & 0 & -p_3(\theta) & -n_3 \\ p_1(\theta) & p_2(\theta) & p_3(\theta) & 0 & -e^+ \\ n_1 & n_2 & n_3 & e^+ & 0 \end{bmatrix}_L,$$

where $p(\theta)=p\cos\theta-p'\sin\theta$. The $\underline{5}$ and $\underline{10}$ containing muons are obtained from these by the replacements $e^+\to\mu^+$, $\nu\to\nu'$, $n\to\lambda$, $\bar{p}\to\bar{p}'$, and $p(\theta)\to p'(\theta)=p'\cos\theta+p\sin\theta$.

Having the representations before us, we answer the obvious questions: Are there anomalies? What Higgs mesons are necessary? What mixing angle is predicted? What new interactions are predicted?

While we already know that the $\mathcal{F}$ subgroup is free of anomalies for the representation we have chosen, the full unifying group might not be. But it is! Remarkably, the $\underline{5}^*$ and $\underline{10}$ have equal and opposite anomalies: Our theory is entirely anomaly free. Indeed, SU(5) is the only group of any rank with a thirty-dimensional, anomaly-free representation with the correct $\mathcal{F}$ content.

Two irreducible representations of Higgs mesons are needed. We need a multiplet with a very large vacuum expectation value to break the SU(5) symmetry down to $\mathcal{F}$. This is done most simply with 24 real scalar-meson fields transforming like the adjoint representation. It is the analog to the superstrong breaking discussed by Weinberg in his treatment of $SU(3)\otimes SU(3)$.[11] All the vector bosons except the twelve associated with generators of $\mathcal{F}$ develop superheavy masses and can hopefully be neglected. We also need Higgs mesons to give mass to the fermions and the weak-interaction intermediaries. For the most general zeroth-order mass matrix consistent with exact color SU(3) symmetry, we need five complex scalar-meson fields transforming like the fundamental representation and 45 complex scalar-meson fields transforming like the $\underline{45}$ contained in $\underline{5}^*\times\underline{10}$. If only the $\underline{5}$ is present, the p and p' masses and the Cabibbo angle are arbitrary, but the other masses satisfy the relations $m_n=m_e$ and $m_\lambda=m_\mu$. Does this mean that the muon-electron mass splitting has the same origin as SU(3) breaking?

For the mixing angle, the theory predicts $\sin^2\theta_w=\frac{3}{8}$.

Finally we come to a discussion of superweak interactions and SU(3)-colored superheavy vector bosons. In addition to mediating such bizarre interactions as $K^0\to\mu^+e^-$, they make the proton unstable. For instance, there is a superheavy colored vector boson which causes the virtual transitions $p_1+p_2\to W\to\bar{n}_3+e^+$. Exchange of this vector boson contributes directly to the decay $p\to\pi^0+e^+$. Since the proton is rather stable,[15] this vector boson must be very massive.[16] The Higgs mesons can also mediate proton decay, and must also be very massive.

From simple beginnings we have constructed the unique simple theory. It makes just one easily testable prediction, $\sin^2\theta_w=\frac{3}{8}$. It also predicts that the proton decays—but with an unknown and adjustable rate. More theoretical work is needed to determine whether the idea of infrared slavery, necessary for our unification, actually makes sense.

---

*Junior Fellow, Harvard University, Society of Fellows. Work supported in part by the U.S. Air Force Office of Scientific Research under Contract No. F44620-70-C-0030 and by the National Science Foundation under Grant No. GP-30819X.

[1]S. L. Glashow, Nucl. Phys. 22, 579 (1961); S. Weinberg, Phys. Rev. Lett. 19, 1264 (1967); A. Salam, in *Proceedings of the Eigth Nobel Symposium, on Elementary Particle Theory, Relativistic Groups, and Analyticity, Stockholm, Sweden, 1968*, edited by N. Svartholm (Almqvist and Wikell, Stockholm, 1968).

VOLUME 32, NUMBER 8 PHYSICAL REVIEW LETTERS 25 FEBRUARY 1974

[2]S. L. Glashow, J. Iliopoulos, and L. Maiani, Phys. Rev. D 2, 1285 (1970).

[3]C. Bouchiat, J. Iliopoulos, and Ph. Meyer, Phys. Lett. 38B, 519 (1972).

[4]M. Gell-Mann, Acta Phys. Austr. Suppl. IX, 733 (1972).

[5]The same statements could be made about a three-quartet model with integral charged quarks, in the spirit of M. Y. Han and Y. Nambu, Phys. Rev. 139, B1006 (1965). In such a scheme, electric charge does not commute with color SU(3).

[6]J. C. Pati and A. Salam, Phys. Rev. D 8, 1240 (1973), and Phys. Rev. Lett. 31, 661 (1973).

[7]S. Weinberg, Phys. Rev. D 8, 605 (1973), and Phys. Rev. Lett. 31, 494 (1973).

[8]O. W. Greenberg, Phys. Rev. Lett. 13, 598 (1964).

[9]H. D. Politzer, Phys. Rev. Lett. 30, 1346 (1973); D. J. Gross and F. Wilczek, Phys. Rev. Lett. 30, 1343 (1973); T. Appelquist and H. Georgi, Phys. Rev. D 8, 4000 (1973); A. Zee, Phys. Rev. D (to be published); H. Georgi and H. D. Politzer, Phys. Rev. D (to be published); D. J. Gross and F. Wilczek, Phys. Rev. D 8, 3633 (1973), and Phys. Rev. D (to be published).

[10]In a general gauge theory, the Lie algebra of the gauge group is a direct sum of a semisimple and an Abelian Lie algebra. Only if the Abelian term is absent—as in a unified theory—is the gauge group necessarily compact, and charge necessarily quantized. Assume that electric charge Q were not quantized in such a theory, i.e., that its eigenvalues were not commensurate. The topological closure of the one-parameter subgroup $\{\exp(i\alpha Q)\}$ would be a compact Abelian Lie group with at least two parameters. Let its Lie algebra A be spanned by $Q, Q_1, \ldots, Q_n$, where $n > 0$. Because the gauge group is compact, its Lie algebra contains A and the Q_i are associated with gauge fields other than the photon. Because $\exp(i\alpha Q)$ is an unbroken local symmetry, the Q_i generate unbroken local symmetries and their associated gauge fields are massless. Since there is only one massless gauge field, we conclude that Q is quantized.

[11]S. Weinberg, Phys. Rev. D 5, 1962 (1972); H. Georgi and S. L. Glashow, Phys. Rev. D 7, 2457 (1973).

[12]See, for instance, T. P. Cheng, E. Eichten, and L.-F. Li, SLAC Report No. SLAC-PUB-1340, 1973 (unpublished).

[13]See the second paper in Ref. 7 and the sixth paper in Ref. 9.

[14]S. Weinberg, to be published.

[15]H. S. Gurr, W. R. Kropp, F. Reines, and B. Meyer, Phys. Rev. 158, 1321 (1967).

[16]A naive calculation indicates that the vector boson mass must be greater than 10^{15} GeV $\simeq 10^{-9}$ g! Let the reader who finds this hard to swallow double the number of fermion states and put quarks and leptons in different (but equivalent) thirty-dimensional representations. He must introduce both weakly interacting quarks and strongly interacting leptons. Now quark number is conserved modulo two and the proton is stable. The deuteron decays via the exchange of four superweak vector bosons, but this is not a serious problem.

VOLUME 33, NUMBER 7 PHYSICAL REVIEW LETTERS 12 AUGUST 1974

Hierarchy of Interactions in Unified Gauge Theories*

H. Georgi,† H. R. Quinn, and S. Weinberg
Lyman Laboratory of Physics, Harvard University, Cambridge, Massachusetts 02138
(Received 15 May 1974)

We present a general formalism for calculating the renormalization effects which make strong interactions strong in simple gauge theories of strong, electromagnetic, and weak interactions. In an SU(5) model the superheavy gauge bosons arising in the spontaneous breakdown to observed interactions have mass perhaps as large as 10^{17} GeV, almost the Planck mass. Mixing-angle predictions are substantially modified.

The scaling observed in deep inelastic electron scattering suggests that what are usually called the strong interactions are not so strong at high energies. Asymptotically free gauge theories of the strong interactions[1] provide a possible explanation: The gluon coupling constant $g(\mu)$ (defined as the value of a three-gluon or gluon-fermion-fermion vertex with momenta characterized by a mass μ) is small when μ is several GeV or larger, but becomes large when μ is small, through the piling up of the logarithms encountered in perturbation theory. In one recent calculation[2] a fit was found for a gauge coupling [in a color SU(3) model][3] with $g^2(\mu)/4\pi \simeq 0.1$ when $\mu \sim 2$ GeV.

If $g(\mu)$ is small when μ is large, then perhaps the strong gauge coupling at some large fundamental mass is of the same order as the couplings in gauge theories of the weak and electromagnetic interactions.[4] Georgi and Glashow[5] have recently gone one step farther, and proposed a model based on the *simple* gauge group SU(5), in which there naturally appears only one free gauge coupling. In their model, SU(5) suffers a spontaneous breakdown to the gauge subgroups SU(3) and SU(2)⊗U(1), which are associated respectively with the strong[3] and the weak and electromagnetic[6] interactions. In order to suppress unobserved interactions, Georgi and Glashow made the necessary assumption[7] that some vector bosons are superheavy.

We find the notion of a simple gauge group uniting strong, weak, and electromagnetic interactions extraordinarily attractive. However, as emphasized by Georgi and Glashow, the success of any such scheme hinges on an understanding of the effects which produce the obvious disparity in strength between the strong and the weak and electromagnetic interactions at ordinary energies. We therefore wish to present in this paper a general formalism for the calculation of such effects. This will lead us to an estimate of the mass of the superheavy gauge bosons. Where a specific model of the gauge groups of the observed interactions is needed as an example, we shall assume that the strong and the weak and electromagnetic interactions are described by color SU(3)[3] and by SU(2)⊗U(1), respectively, and where a specific example of a unifying simple gauge group is needed, we shall use SU(5).

If we neglect all renormalization effects, the embedding of the gauge groups G_i of the observed interactions in a larger simple group G imposes a relation among their coupling constants. We

VOLUME 33, NUMBER 7 PHYSICAL REVIEW LETTERS 12 AUGUST 1974

normalize the generators T_α of G so that in any representation D of G we have

$$\mathrm{Tr}(T_\alpha T_\beta) = N_D \delta_{\alpha\beta}, \tag{1}$$

where N_D may depend on the representation but not on α and β. We use the same normalization conventions for the gauge groups of the observed interactions. Then invariance under G implies that the coupling constants g_G, g_3, g_2, and g_1 associated with the group G and the subgroups SU(3), SU(2), and U(1), respectively, are equal. The usual SU(2) and U(1) coupling constants[6] may be identified as

$$g = g_2, \quad g' = g_1/C, \tag{2}$$

where C is a constant entering the relation between the charge Q and the SU(2) and U(1) generators $\vec{T}$ and T_0, normalized according to Eq. (1):

$$Q = T_3 - CT_0. \tag{3}$$

The weak mixing angle[6] is then given by

$$\sin^2\theta = e^2/g^2 = g'^2/(g^2 + g'^2) = (1 + C^2)^{-1}. \tag{4}$$

In any representation of G, reducible or irreducible,

$$\mathrm{Tr}(Q^2) = (1 + C^2)\mathrm{Tr}(T_3^2). \tag{5}$$

If we take our representation to consist of the left-handed states of three quartets of colored quarks, three antiquark quartets, and ν_e, ν_μ, e^-, e^+, μ^-, μ^+, then there are eight SU(2) doublets, and so $\mathrm{Tr}(T_3^2) = 4$, while $\mathrm{Tr}(Q^2) = \frac{32}{3}$, so that

$$C^2 = \tfrac{5}{3}, \quad \sin^2\theta = \tfrac{3}{8}. \tag{6}$$

This is the case for the SU(5) model.[5] We shall leave C arbitrary in what follows, and will find that the choice of the simple unifying group G enters the calculation only through the single parameter C.

Now let us see how to take renormalization effects into account. The gauge couplings are functions of the momentum scale μ, and the above relations among gauge couplings really only apply when μ is much larger than the superheavy boson masses, where the breaking of G may be neglected. However, the observed values of the gauge couplings refer to much smaller values of μ, of the order of the W and Z masses, or even smaller. The problem is to bridge the gap between superlarge values of μ, where G imposes relations among the gauge couplings, and ordinary values of μ, where the gauge couplings are observed.

In order to accomplish this, we make use of the theorem[8] that all matrix elements involving only "ordinary" external particles with momenta and masses much less than all superheavy masses may be calculated in an effective renormalizable field theory, which is just the original field theory with all superheavy particles omitted, but with coupling constants that may depend on the superheavy masses. All other effects of the superheavy particles are suppressed by factors of an ordinary mass divided by a superheavy mass.

When μ is large compared with all ordinary masses but small compared with all superheavy masses, the μ dependence of the couplings is governed by a renormalization-group equation,[9]

$$\mu \frac{d}{d\mu} g_i(\mu) = \beta_i(g_i(\mu)), \tag{7}$$

with β_i calculated in the effective field theory based on the "observed" gauge group G_0. If all $g(\mu)$ are small, then β_i depends only on g_i, with[10]

$$\beta_i(g(\mu)) \simeq b_i g_i^3(\mu) \text{ for } |g_i| \ll 1, \tag{8}$$

so that

$$g_i^{-2}(\mu) \simeq \text{const} - 2b_i \ln\mu. \tag{9}$$

The integration constants are determined by the underlying simple group G. Specifically, if we suppose that all superheavy gauge bosons have masses of the order of some typical superheavy mass M, and if we take μ to be of the order of but somewhat smaller than M, then the $g_i(\mu)$ may be calculated perturbatively in the simple gauge theory based on G, and as long as $g_i(M)$ is sufficiently small, each gauge coupling will be essentially given by its group-theoretic value g_G neglecting all renormalizations. Thus Eq. (9) gives

$$g_i^{-2}(\mu) = g_G^{-2}(M) + 2b_i \ln(M/\mu) \tag{10}$$

for $\mu \leq M$.

The gauge coupling constants g_{i_0} observed in present experiments are essentially given by the values of the $g_i(\mu)$ when μ is some "ordinary" mass m, of the order of 10 GeV. Since all these couplings are small and therefore slowly varying in the range of interest, our result is not particularly sensitive to the value of the "ordinary" mass at which we choose to study the couplings.

Let us now specifically assume that the "observed" gauge group is SU(3)⊗SU(2)⊗U(1), but for the moment leave open the choice of the group G. Choosing convenient linear combinations of

VOLUME 33, NUMBER 7 PHYSICAL REVIEW LETTERS 12 AUGUST 1974

(10), we have

$$\frac{C^2}{g_{1o}^{\;2}}+\frac{1}{g_{2o}^{\;2}}-\frac{(1+C^2)}{g_{3o}^{\;2}}=\frac{1}{e^2}-\frac{(1+C^2)}{g_{3o}^{\;2}}$$

$$=2\ln\left(\frac{M}{m}\right)[b_1C^2+b_2-b_3(1+C^2)], \qquad (11)$$

$$C^2(g_{1o}^{\;-2}-g_{2o}^{\;-2})=e^{-2}[1-(1+C^2)\sin^2\theta]$$

$$=2C^2(b_1-b_2)\ln(M/m). \qquad (12)$$

To calculate the b's, we note that any multiplet of particles forming a representation of G does not contribute at all to the b's if *all* particles are superheavy, while it contributes equally to all b's if *no* particles in the multiplet are superheavy, and therefore in either case has no effect in Eqs. (11) and (12). We shall assume that the only multiplet which contains *both* ordinary and superheavy particles is the gauge multiplet itself, in which case[1]

$$b_2=-\tfrac{22}{3}(4\pi)^{-2}+b_1, \quad b_3=-11(4\pi)^{-2}+b_1; \qquad (13)$$

so (11) and (12) give

$$\ln\left(\frac{M}{m}\right)=\frac{3(4\pi)^2}{22(1+3C^2)}\left(\frac{1}{e^2}-\frac{(1+C^2)}{g_{3o}^{\;2}}\right), \qquad (14)$$

$$\sin^2\theta=(1+3C^2)^{-1}(1+2C^2e^2/g_{3o}^{\;2}). \qquad (15)$$

For $C^2=\frac{5}{3}$ [the SU(5) value], $m=10$ GeV, and reasonable values of $g_{3o}^{\;2}/4\pi$, we obtain the results displayed in the following table:

$g_{3o}^{\;2}/4\pi$	M (GeV)	$\sin^2\theta$
0.5	2×10^{17}	0.175
0.2	2×10^{16}	0.187
0.1	5×10^{14}	0.207
0.05	2×10^{11}	0.248

It is intriguing that we are led to contemplate elementary particle masses as high as 2×10^{17} GeV, of about the same order of magnitude as the Planck mass, $G^{-1/2}=1.2206\times10^{19}$ GeV. Perhaps gravitation has something to do with the superstrong spontaneous symmetry breaking, or perhaps the spontaneous breakdown of the simple gauge group has something to do with setting the scale of the gravitational interaction.

Equation (15) predicts lower values for $\sin\theta$ than does Eq. (4). While the available data favor the higher value, they are rather preliminary, and the strong constraints on $\sin^2\theta$ follow only if the Z mass is assumed to satisfy the relation $M_W^{\;2}=M_Z^{\;2}\cos^2\theta_W$, which depends on the Higgs structure of the model.[11] If this relation is abandoned, $\sin^2\theta_W$ must be inferred from the ratio of the ratio of neutral- to charged-current events seen in neutrino scattering to that seen in antineutrino scattering.[11]

In the SU(5) example, it is not necessarily true that all effects of order m/M are negligible, because some of the superheavy vector bosons mediate proton decay into lepton plus pions. Since the proton is otherwise stable, such very small effects may be observable. Calculation of the proton lifetime involves details of the strong interactions at small momenta, but we can give an order-of-magnitude estimate on dimensional grounds. The lifetime must be proportional to M^4 and so it must approximately equal $M^4/m_p^{\;5}$. Taking $M=5\times10^{15}$ GeV, for example, gives a proton lifetime of about 6×10^{31} yr. The present experimental lower limit is 10^{30} yr.[12] The observation of proton decay with a lifetime of this order of magnitude would be a startling confirmation of the ideas discussed here.

Before concluding, we emphasize again the approximation which went into the derivation of Eqs. (14) and (15). We have idealized the two transition regions: the region in momentum scale around M where the three coupling constants are merging into one, and the region from m into the timelike domain where we actually measure e^2. The corrections to (14) and (15) due to changes of the coupling constants in these regions can be calculated using perturbation theory in the relevant coupling constants. The corrections for the second region are electromagnetic and therefore small (g_{3o} can in principle be measured directly in the spacelike region through the observation of logarithmic violations of scaling in electroproduction). The corrections from the first region will be small if $g_G(M)$ is small. To calculate $g_G(M)$ we need to know the fermion and scalar-meson content of the theory. For the SU(5) model with $M=5\times10^{15}$ GeV (see the table), $g_G^{\;2}(M)/4\pi=(48\pm1)^{-1}$.

We have also assumed that the lowest-order form for β_i, Eq. (8), is valid down to $\mu=m$. Next-order corrections to β have been calculated by Belavin and Migdal.[13] We use their results and find the ratio of the correction to the lowest-order value in the SU(5) theory to be about $0.6g^2/4\pi$. Such corrections are obviously only relevant for $g_3(\mu)$, and even for $g_{3o}^{\;2}/4\pi=0.5$, the largest value used in the table, the correction is only 30%.

Finally we want to emphasize what seems to us to be the most disturbing feature of the class of models discussed here, that is, the existence of

VOLUME 33, NUMBER 7 PHYSICAL REVIEW LETTERS 12 AUGUST 1974

two stages of spontaneous symmetry breaking characterized by radically different mass scales. In the context of the conventional Higgs mechanism, we can find no natural explanation of the enormous ratio of superheavy mass to ordinary mass. We have nothing quantitative to say about this mystery, but the following speculation seems attractive to us. Suppose that only superstrong breaking takes place via the Higgs mechanism. There is only one mass scale in the theory and it is superheavy. All of the scalar mesons are either superheavy or Goldstone bosons (note that this obviates the difficulties associated with superlarge trilinear couplings among ordinary-mass scalars). Well below the superheavy mass scale the theory is an effective SU(3)⊗SU(2)⊗U(1) theory containing only gauge fields and fermions. The next stage of symmetry breaking is dynamical and hence nonperturbative. The mass scale associated with this stage is the mass at which $g_3(\mu)$ gets large enough that nonperturbative effects become important.

We are grateful for discussions with T. Appelquist, S. Coleman, S. L. Glashow, and H. D. Politzer.

---

*Work supported in part by the National Science Foundation under Grant No. GP40397X.

†Junior Fellow, Harvard University Society of Fellows.

[1]H. D. Politzer, Phys. Rev. Lett. 30, 1346 (1973); D. J. Gross and F. Wilczek, Phys. Rev. Lett. 30, 1343 (1973).

[2]H. D. Politzer, to be published. Of course, given the available data, any such estimate is necessarily very crude.

[3]W. Bardeen, H. Fritsch, and M. Gell-Mann, in *Scale and Conformal Invariance in Hadron Physics*, edited by R. Gatto (Wiley, New York, 1973), p. 139. Also see R. H. Dalitz, in *High Energy Physics*, edited by C. DeWitt and M. Jacob (Gordon and Breach, New York, 1966), p. 287.

[4]S. Weinberg, J. Phys. (Paris), Colloq. 34, C1–45 (1973), and to be published.

[5]H. Georgi and S. L. Glashow, Phys. Rev. Lett. 32, 438 (1974).

[6]S. Weinberg, Phys. Rev. Lett. 19, 1264 (1967); A. Salam, in *Elementary Particle Physics*, edited by N. Svartholm (Almquist and Wiksels, Stockholm, 1968), p. 367.

[7]S. Weinberg, Phys. Rev. D 5, 1962 (1972).

[8]T. Appelquist and J. Carrazone, to be published. A different proof of the same result for graphs including only a single superheavy line is given by S. Weinberg, Phys. Rev. D 8, 605, 4482 (1973). This theorem and hence our entire discussion could be invalidated if there were superlarge trilinear couplings among scalar fields. The survival of any light scalars already requires that certain scalar self-couplings in the symmetric theory were made extremely small. It appears that the problem of large trilinear couplings can always be avoided by some device. We will discuss later in this paper a possible point of view which avoids this unattractive situation altogether.

[9]M. Gell-Mann and F. E. Low, Phys. Rev. 95, 1300 (1954); C. G. Callan, Phys. Rev. D 2, 1541 (1970); K. Symanzik, Commun. Math. Phys. 18, 227 (1970).

[10]It was H. D. Politzer who pointed out to us that perturbation theory could be used down to the energy where the strong interactions begin to be much stronger than the weak and electromagnetic interactions, so that it is not necessary here to worry about the region where the strong interactions are really strong.

[11]A. De Rújula, H. Georgi, S. L. Glashow, and H. Quinn, to be published.

[12]F. Reines and M. F. Crouch, Phys. Rev. Lett. 32, 493 (1974).

[13]A. A. Belavin and A. A. Migdal, to be published. We learned recently that the same calculation has been performed by W. E. Caswell [Phys. Rev. Lett. 33, 244 (1974)]. His result differs slightly from that given by Belavin and Migdal, but the difference is insignificant as far as our estimate is concerned.

Volume 59B, number 3 PHYSICS LETTERS 10 November 1975

ARE QUARKS COMPOSITE?

J.C. PATI
Department of Physics and Astronomy, University of Maryland, College Park, Maryland, USA

A. SALAM
International Centre for Theoretical Physics, Trieste, Italy and Imperial College, London, England

and

J. STRATHDEE
International Centre for Theoretical Physics, Trieste, Italy

Received 15 September 1975

It is suggested that quarks and leptons are composites of still more fundamental PRE-entities.

An elegant economy was introduced in hadron physics when Gell-Mann and Zweig invented the quark. Each quark represented one "valency attribute": *I*-spin up, *I*-spin down and strangeness. All hadrons were assumed to be quark composites. The original three valency attributes have since presumably increased to four with the inclusion of charm. Another extension, the inclusion of three colours was needed principally to resolve a spin-statistics dilemma. Most theorists thus believe in twelve fundamental quarks representing seven fundamental attributes (four valencies and three colours). Another doubling, to twenty-four quarks, has been suggested recently to accommodate the "mirror" quantum number which may be needed to give a description of J/ψ particles. If lepton number is counted as a fourth colour [‡1] one would arrive at the need for a thirty-two component fundamental quark of which all matter is made.

Clearly the time has come to consider alternative ways of accommodating the fundamental attributes. It could prove useful to regard the quarks themselves as composite structures and our purpose here is to sketch a model based on entities which may be more basic than quarks. These entities we shall call PRE's, and of them the thirty-two quarks are supposed to be made [‡1]. To represent the PRE's a pair of quartets is needed. One, the *valency quartet* $\mathcal{Q} = (\mathrm{pn}\lambda\chi)$, serves to carry the usual SU(3) quantum numbers and charm. The other, the *colour quartet*, $\mathcal{C} = (\mathrm{abcd})$, carries the quantum numbers of SU(3) colour and lepton number. There are two possible variants:

(A) $\mathcal{Q}$ is a fermion and $\mathcal{C}$ a boson, (B) $\mathcal{Q}$ is a boson and $\mathcal{C}$ a fermion.

The sixteen quark states [‡2], $\mathcal{Q}\bar{\mathcal{C}}$, are two-body composites and, of course, fermionic. Neither of these schemes lend themselves to an SU(6) type of classification at the PRE level (though this is perhaps not a very serious objection). However, scheme (B) cannot be used for a gauge theory of weak interactions and we shall discard it.

With the introduction of a neutral PRE singlet fermion, $\mathcal{S}$, it is possible to envision two other variants in which the quarks appear as three-particle composites [‡2], $\mathcal{Q}\bar{\mathcal{C}}\mathcal{S}$:

(C) $\mathcal{Q}$ and $\mathcal{C}$ are bosons, (D) $\mathcal{Q}$ and $\mathcal{C}$ are fermions.

‡1 The idea of PRE's was motivated by Pati and Salam [1,2]. The same idea has been independently considered by Greenberg [3], who gives references to earlier work.

‡2 To incorporate the "mirror" (or "heaviness") quantum number introduce one additional, electically neutral singlet PRE, $\mathcal{S}'$. In this case the "mirror" (heavy) quarks are $\mathcal{Q}\bar{\mathcal{C}}\mathcal{S}'$ composites.

Scheme (C) shares the difficulties of (B). Thus, schemes (A) and (D) are the two variants one may seriously consider. Below we present a gauge model for (D) which, with minor modifications, can be adapted for (A) as well.

The underlying global symmetry is $(SU(4)_L \times SU(4)_R \times U(1))_{valence} \times (SU(4) \times U(1))_{colour}$. To avoid problems with anomalies, we shall restrict the valence (local) symmetry to $SU(2)_L \times SU(2)_R \times U(1)$. The presence of two U(1)'s serves two ends. Firstly, these contribute to the PRE charges which can therefore be assigned integer values [‡3]. Secondly, the U(1)'s provide gauge couplings for the neutral singlet $\mathcal{S}$ with itself and with $\mathcal{Q}$'s and $\mathcal{C}$'s so that the composites, $\mathcal{Q}\bar{\mathcal{C}}\mathcal{S}$, can appear as bound states. We assign particles as follows:

$$\mathcal{Q}_L = (2{+}2,1,1)_{1,0}\,, \quad \mathcal{Q}_R = (1,2{+}2,1)_{1,0}\,, \quad \mathcal{C} = (1,1,4)_{0,1}\,, \quad \mathcal{S} = (1,1,1)_{-1,1}\,. \tag{1}$$

(Here the $U(1)_{valence} \times U(1)_{colour}$ assignments are shown as subscripts.) The charge operator

$$Q_{electric} = I_{3L} + I_{3R} + F'_3 + \frac{1}{\sqrt{3}} F'_8 - \sqrt{\tfrac{2}{3}}\, F'_{15} - \tfrac{1}{2} I_0 - \tfrac{1}{2} I'_0 \tag{2}$$

takes the form diag(0,−1,−1,0) on the quartets and of course vanishes on the singlet. Notice that the $U(1) \times U(1)$ quantum numbers assigned to $\mathcal{S}$ are opposite to those on $\mathcal{Q}$ and $\bar{\mathcal{C}}$ and so would yield attractive gluon forces in the composite $\mathcal{Q}\mathcal{C}\mathcal{S}$.

A Lagrangian model for the PRE's with couplings mediated by gauge fields (whose masses are generated by a Higgs mechanism) is easily constructed. The valence symmetry, $SU(2)_L \times SU(2)_R \times U(1)$ is associated with two triplets, W_L, W_R, and a singlet, T, while the colour symmetry $SU(4) \times U(1)$ is associated with a 15-fold V, and a singlet T'. To generate masses for these vector fields we shall take the Higgs-Kibble set:

$$A = (4,\bar{4},1)_{0,0}\,, \quad B = (1,4,\bar{4})_{1,-1}\,, \quad C = (\bar{4},1,4)_{-1,1}\,, \quad D = (1,1,4)_{0,1}\,. \tag{3}$$

The subset consisting of A, B and C has been analysed previously (on the assumption that the dominant terms in the potential are invariant under the full global symmetry $(U(4)_L \times U(4)_R)_{valence} \times U(4)_{colour}$). Here, because of the singlets T and T' it is necessary to include a fourth multiplet, D (to which we assign the quantum numbers of the colour PRE), in order that the only residual symmetry shall be the electromagnetic U(1).

The Higgs potential can be taken in such a way as to force the vacuum expectation values into the forms

$$\langle A\rangle = \mathrm{diag}(a_1, a_1, a_1, a_4)\,, \quad \langle B\rangle = \mathrm{diag}(0,0,0,b_4)\,, \quad \langle C\rangle = \mathrm{diag}(c_1, c_1, c_1, c_4)\,, \quad \langle D\rangle = (0,0,0,d). \tag{4}$$

Physical considerations (which apply as much to the model of ref. [1] as to the present one) give the scale $b \sim 10^4 - 10^5$ GeV; $a \sim 300$ GeV; $c \sim 1$ GeV. The mass term for Fermi particles is taken in the form $m_c \bar{\mathcal{C}}\mathcal{C} + m_s \bar{\mathcal{S}}\mathcal{S} + K\bar{\mathcal{Q}}_L A^+ \mathcal{Q}_R$ + h.c. (The couplings $\bar{\mathcal{Q}}_R B\, \mathcal{C}_L$ and $\bar{\mathcal{Q}}_L C^\dagger\, \mathcal{C}_R$ may be excluded by the imposition of a discrete symmetry $\mathcal{Q} \to -\mathcal{Q}$, $\mathcal{C} \to \mathcal{C}$.)

Of particular importance is the vector mass matrix which determines the complexion of the gauge interactions. Except for the parts involving the singlets T and T' (whose presence necessitates the introduction of D) this matrix has been analysed in ref. [1]. The contribution of D will cause some not very significant modifications in the masses of the charged vectors, but the neutral (diagonal) components will be affected more drastically. These are given by

$$\frac{g^2}{4}(3a_1^2 + a_4^2)(W_L - W_R)^2 + \frac{b_4^2}{4}\left(gW_R + f\sqrt{\frac{3}{2}}V_{15} + hT - h'T'\right)^2 + \frac{c_4^2}{4}\left(gW_L + f\sqrt{\frac{3}{2}}V_{15} + hT - h'T'\right)^2$$
$$+ \frac{c_1^2}{4}\left(-gW_L + f\left(V_3 + \frac{V_8}{\sqrt{3}} + \frac{V_{15}}{\sqrt{6}}\right) - hT + h'T'\right)^2 + \frac{c_1^2}{4}\left(gW_L + f\left(-V_3 + \frac{V_8}{\sqrt{3}} + \frac{V_{15}}{\sqrt{6}}\right) - hT + h'T'\right)^2$$
$$+ \frac{c_1^2}{4}\left(gW_L + f\left(-\frac{2V_8}{\sqrt{3}} + \frac{V_{15}}{\sqrt{6}}\right) - hT + h'T'\right)^2 + \frac{d^2}{4}\left(f\sqrt{\frac{3}{2}}V_{15} + h'T'\right) + \frac{\mu^2}{2}(hT - h'T')^2\,. \tag{5}$$

‡3 This is admittedly a prejudice. If the U(1)'s are discarded, then the quartet PRE's would carry electric charges ± 1/2. Note that the total U(1) charge on quarks or leptons is zero.

Volume 59B, number 3 PHYSICS LETTERS 10 November 1975

The last term is the only non-Higgs contribution consistent with both renormalizability and electromagnetic gauge invariance.

The photon is expressed by the exact formula

$$\frac{1}{e}A = \frac{1}{g}(W_L + W_R) + \frac{1}{f}\left(V_3 + \frac{1}{\sqrt{3}}V_8 - \sqrt{\frac{2}{3}}V_{15}\right) + \frac{1}{h}T + \frac{1}{h'}T', \tag{6}$$

but the other seven states are thoroughly mixed and we shall attempt only a very approximate diagonalization (for a specially favourable sequence of parameters). We assume that g is small relative [‡4] to f, h, h' and we shall neglect terms of order g/f, g/h and g/h'. Further, we assume that $b > \mu > d > c$ and neglect terms of order $(\mu/b)^2$, $(d/\mu)^2$, $(c/d)^2$. In this approximation one can treat the subsystems $\{W_L, W_R\}$, $\{V_3, V_8\}$ and $\{V_{15}, T, T'\}$ independently [‡5]. One finds

$$X_1 = \frac{f\sqrt{\frac{3}{2}}V_{15} + hT - h'T'}{\sqrt{\frac{3}{2}f^2 + h^2 + h'^2}}, \qquad X_2 = \frac{-(h^2+h'^2)V_{15} + \sqrt{\frac{3}{2}}f(hT - h'T')}{\sqrt{(h^2+h'^2)(\frac{3}{2}f^2 + h^2 + h'^2)}}, \qquad X_3 = \frac{h'T + hT'}{\sqrt{h^2+h'^2}}, \tag{7}$$

with the respective masses

$$m(X_1) = \frac{1}{\sqrt{2}}b_4\sqrt{\tfrac{3}{2}f^2 + h^2 + h'^2}, \qquad m(X_2) = \sqrt{\tfrac{3}{2}}\,\mu\sqrt{\frac{h^2+h'^2}{\frac{3}{2}f^2 + h^2 + h'^2}}, \qquad m(X_3) = \frac{1}{\sqrt{2}}d\frac{hh'}{\sqrt{h^2+h'^2}}$$

while the octet $V(8)$ has the mass fc_1.

The static interaction between the PRE's mediated by V_{15}, T and T' is dominated by the X_3 pole whose contribution to the vector propagators is summarized by the matrix

$$\begin{pmatrix} (VV) & (VT) & (VT') \\ (TV) & (TT) & (TT') \\ (T'V) & (T'T) & (T'T') \end{pmatrix} \sim \frac{\mathrm{i}}{k^2 - m(X_3)^2}\,\frac{1}{h^2+h'^2}\begin{pmatrix} 0 & 0 & 0 \\ 0 & h'^2 & hh' \\ 0 & hh' & h^2 \end{pmatrix}. \tag{8}$$

The important point is that the inter-PRE forces have a range $m(X_3)^{-1} \sim 1/hd$ which is considerably shorter than the range, $1/fc_1$, of the strong inter-$\mathcal{C}$ forces mediated by the colour octet, which may perhaps be identifiable with the J/ψ particles with masses in the range 3–5 GeV. The model indicates that leptons will experience strong interactions at energies of order $m(X_3) \approx d(h^2h'^2/(2h^2 + 2h'^2))^{1/2}$. Thus d must be chosen $\gtrsim a \gtrsim 300$ GeV.

The picture which emerges is one of very short range colour and valence singlet inter-PRE forces which give rise to composite quark and lepton states. The longer range inter-quark forces, however, are dominated by exchanges of the colour octet.

The above considerations have all been qualitative. The problem of computing bound states is of course non-trivial even when feasible. The hardest among these problems is the question of zero-mass leptonic composites, i.e. neutrinos. While it is not impossible that such states could arise due to fortuitous relationships among the various masses and couplings in the Lagrangian, we do not feel that such a eventuality is plausible. Rather, we should like to view the neutrinos (like the photon) as rather special "composites" whose existence should be guaranteed by a symmetry principle. In the case of the photon, the relevant symmetry is of course gauge invariance. For the neutrino it may be possible to incorporate supersymmetry (whose spontaneous breakdown leads inevitably to the appearance of zero-mass spinors).

The eight PRE's we have introduced in this paper represent eight internal symmetry attributes (four valencies plus four colours). It is hard at present to conceive of a theory which uses fewer fundamental entities, unless some

We thank Professor C.H. Woo for several helpful discussions on the problem of compositness.

‡4 We are assuming that the two U(1)'s represent strong or medium strong interactions. This ensures that $\mathcal{Q}$, $\bar{\mathcal{C}}$ and $\mathcal{S}$ bind to form quarks and leptons. Thus $g^2/4\pi = 2\alpha < h^2/4\pi, h'^2/4\pi \leqslant f^2/4\pi$.

‡5 In ref. [1] the field V_{15} was denoted by S^0.

Volume 59B, number 3 PHYSICS LETTERS 10 November 1975

of the attributes disappear experimentally. The main point of this note, however, is that each fundamental attribute should be associated with one fundamental PRE-entity.

References

[1] J.C. Pati and A. Salam, Phys. Rev. D10 (1974) 275.
[2] J.C. Pati and A. Salam, ICTP, Trieste, preprint IC/75/106 (to appear in the Proc. Palermo Conf., June 1975).
[3] O.W. Greenberg, University of Maryland, technical report 76-012 (1975).
[4] J.C. Pati and A. Salam, ICTP, Trieste, preprint IC/75/73 (Phys. Letters B, to be published).

NATURALNESS, CHIRAL SYMMETRY, AND SPONTANEOUS CHIRAL SYMMETRY BREAKING

G. 't Hooft

Institute for Theoretical Fysics

Utrecht, The Netherlands

ABSTRACT

A properly called "naturalness" is imposed on gauge theories. It is an order-of-magnitude restriction that must hold at all energy scales μ. To construct models with complete naturalness for elementary particles one needs more types of confining gauge theories besides quantum chromodynamics. We propose a search program for models with improved naturalness and concentrate on the possibility that presently elementary fermions can be considered as composite. Chiral symmetry must then be responsible for the masslessness of these fermions. Thus we search for QCD-like models where chiral symmetry is not or only partly broken spontaneously. They are restricted by index relations that often cannot be satisfied by other than unphysical fractional indices. This difficulty made the author's own search unsuccessful so far. As a by-product we find yet another reason why in ordinary QCD chiral symmetry must be broken spontaneously.

III1. INTRODUCTION

The concept of causality requires that macroscopic phenomena follow from microscopic equations. Thus the properties of liquids and solids follow from the microscopic properties of molecules and atoms. One may either consider these microscopic properties to have been chosen at random by Nature, or attempt to deduce these from even more fundamental equations at still smaller length and time scales. In either case, it is unlikely that the microscopic equations contain various free parameters that are carefully adjusted by Nature to give cancelling effects such that the macroscopic systems have some special properties. This is a philosophy which we would like to apply to the unified gauge theories: the effective interactions at a large length scale, corresponding to a low energy scale μ_1, should follow from the properties at a much smaller length scale, or higher energy scale μ_2, without the requirement that various different parameters at the energy scale μ_2 match with an accuracy of the order of μ_1/μ_2. That would be unnatural. On the other hand, if at the energy scale μ_2 some parameters would be very small, say

$$\alpha(\mu_2) = \mathcal{O}(\mu_1/\mu_2) , \qquad \text{(III1)}$$

then this may still be natural, provided that this property would not be spoilt by any higher order effects. We now conjecture that the following dogma should be followed:

- at any energy scale μ, a physical parameter or set of physical parameters $\alpha_i(\mu)$ is allowed to be very small only if the replacement $\alpha_i(\mu) = 0$ would increase the symmetry of the system. -

In what follows this is what we mean by naturalness. It is clearly a weaker requirement than that of P. Dirac[1] who insists on having no small numbers at all. It is what one expects if at any mass scale $\mu > \mu_0$ some ununderstood theory with strong interactions determines a spectrum of particles with various good or bad symmetry properties. If at $\mu = \mu_0$ certain parameters come out to be small, say 10^{-5}, then that cannot be an accident; it must be the consequence of a near symmetry.

For instance, at a mass scale

$\mu = 50$ GeV,

the electron mass m_e is 10^{-5}. This is a small parameter. It is acceptable because $m_e = 0$ would imply an additional chiral symmetry corresponding to separate conservation of left handed and right handed electron-like leptons. This guarantees that all renormalizations of m_e are proportional to m_e itself. In sects. III2 and III3 we compare naturalness for quantum electrodynamics and ϕ^4 theory.

Gauge coupling constants and other (sets of) interaction constants may be small because putting them equal to zero would turn the gauge bosons or other particles into free particles so that they are separately conserved.

If within a set of small parameters one is several orders of magnitude smaller than another then the smallest must satisfy our "dogma" separately. As we will see, naturalness will put the severest restriction on the occurrence of scalar particles in renormalizable theories. In fact we conjecture that this is the reason why light, weakly interacting scalar particles are not seen.

It is our aim to use naturalness as a new guideline to construct models of elementary particles (sect. III4). In practice naturalness will be lost beyond a certain mass scale μ_0, to be referred to as "Naturalness Breakdown Mass Scale" (NBMS). This simply means that unknown particles with masses beyond that scale are ignored in our model. The NBMS is only defined as an order of magnitude and can be obtained for each renormalizable field theory. For present "unified theories", including the existing grand unified schemes, it is only about 1000 GeV. In sect. 5 we attempt to construct realistic models with an NBMS some orders of magnitude higher.

One parameter in our world is unnatural, according to our definition, already at a very low mass scale ($\mu_0 \sim 10^{-2}$ eV). This is the cosmological constant. Putting it equal to zero does not seem to increase the symmetry. Apparently gravitational effects do not obey naturalness in our formulation. We have nothing to say about this fundamental problem, accept to suggest that *only* gravitational effects violate naturalness. Quantum gravity is not understood anyhow so we exclude it from our naturalness requirements.

On the other hand it is quite remarkable that all other elementary particle interactions have a high degree of naturalness. No unnatural parameters occur in that energy range where our popular field theories could be checked experimentally. We consider this as important evidence in favor of the general hypothesis of naturalness. Pursuing naturalness beyond 1000 GeV will require theories that are immensely complex compared with some of the grand unified schemes.

A remarkable attempt towards a natural theory was made by Dimopoulos and Susskind [2)]. These authors employ various kinds of confining gauge forces to obtain scalar bound states which may substitute the Higgs fields in the conventional schemes. In their model the observed fermions are still considered to be elementary.

Most likely a complete model of this kind has to be constructed step by step. One starts with the experimentally accessible aspects of the Glashow-Weinberg-Salam-Ward model. This model is natural if one restricts oneself to mass-energy scales below 1000 GeV. Beyond 1000 GeV one has to assume, as Dimopoulos and Susskind do, that the Higgs field is actually a fermion-antifermion composite field. Coupling this field to quarks and leptons in order to produce their mass, requires new scalar fields that cause naturalness to break down at 30 TeV or so. Dimopoulos and Susskind speculate further on how to remedy this. To supplement such ideas, we toyed with the idea that (some of) the presently "elementary" fermions may turn out to be bound states of an odd number of fermions when considered beyond 30 TeV. The binding mechanism would be similar to the one that keeps quarks inside the proton. However, the proton is not particularly light compared with the characteristic mass scale cf quantum chromodynamics (QCD). Clearly our idea is only viable if something prevented our "baryons" from obtaining a mass (eventually a small mass may be due to some secondary perturbation).

The proton ows its mass to spontaneous breakdown of chiral symmetry, or so it seems according to a simple, fairly successful model of the mesonic and baryonic states in QCD: the Gell-Mann-Lévy sigma model[3]. Is it possible then that in some variant of QCD chiral symmetry is not spontaneously broken, or only partly, so that at least some chiral symmetry remains in the spectrum of fermionic bound states? In this article we will see that in general in SU(N) binding theories this is not allowed to happen, i.e. chiral symmetry must be broken spontaneously.

III2. NATURALNESS IN QUANTUM ELECTRODYNAMICS

Quantum Electrodynamics as a renormalizable model of electrons (and muons if desired) and photons is an example of a "natural" field theory. The parameters α, m_e (and m_μ) may be small independently. In particular m_e (and m_μ) are very small at large μ. The relevant symmetry here is chiral symmetry, for the electron and the muon separately. We need not be concerned about the Adler-Bell-Jackiw anomaly here because the photon field being Abelian cannot acquire non-trivial topological winding numbers[4].

There is a value of μ where Quantum Electrodynamics ceases to be useful, even as a model. The model is not asymptotically free, so there is an energy scale where all interactions become strong:

$$\mu_o \simeq m_e \exp(6\pi^2/e^2N_f) \ , \qquad \text{(III2)}$$

where N_f is the number of light fermions. If some world would be described by such a theory at low energies, then a replacement of the theory would be necessary at or below energies of order μ_o.

III3. ϕ^4-THEORY

A renormalizable scalar field theory is described by the Lagrangian

$$\mathcal{L} = -\tfrac{1}{2}(\partial_\mu\phi)^2 - \tfrac{1}{2}m^2\phi^2 - \frac{1}{4!}\lambda\phi^4 \ . \qquad \text{(III3)}$$

the interactions become strong at

$$\mu \simeq m \exp(16\pi^2/3\lambda) \ , \qquad \text{(III4)}$$

but is it still natural there?

There are two parameters, λ and m. Of these, λ may be small because $\lambda = o$ would correspond to a non-interacting theory with total number of ϕ particles conserved. But is small m allowed? If we put m = o in the Lagrangian (III3) then the symmetry is not enhanced*). However we can take both m and λ to be small, because if $\lambda = m = o$ we have invariance under

$$\phi(x) \rightarrow \phi(x) + \Lambda \ . \tag{III5}$$

This would be an approximate symmetry of a new underlying theory at energies of order μ_o. Let the symmetry be broken by effects described by a dimensionless parameter ε. Both the mass term and the interaction term in the effective Lagrangian (III3) result from these symmetry breaking effects. Both are expected to be of order ε. Substituting the correct powers of μ_o to account for the dimensions of these parameters we have

$$\begin{aligned} \lambda &= \mathcal{O}(\varepsilon) \ , \\ m^2 &= \mathcal{O}(\varepsilon\mu_o^2) \ . \end{aligned} \tag{III6}$$

Therefore,

$$\mu_o = \mathcal{O}(m/\sqrt{\lambda}) \ . \tag{III7}$$

This value is much lower than eq. (III4). We now turn the argument around: if any "natural" underlying theory is to describe a scalar particle whose *effective* Lagrangian at low energies will be eq. (III3), then its energy scale cannot be given by (III4) but at best by (III7). We say that naturalness breaks down beyond $m/\sqrt{\lambda}$. It must be stressed that these are orders of magnitude. For instance one might prefer to consider λ/π^2 rather than λ to be the relevant parameter. μ_o then has to be multiplied by π. Furthermore, λ could be much smaller than ε because $\lambda = o$ separately also enhances the symmetry. Therefore, apart from factors π, eq. (III7) indicates a maximum value for μ_o.

Another way of looking at the problem of naturalness is by comparing field theory with statistical physics. The parameter m/μ would correspond to $(T-T_c)/T$ in a statistical ensemble. Why would the temperature T chosen by Nature to describe the elementary particles be so close to a critical temperature T_c? If $T_c \neq o$ then T may not be close to T_c just by accident.

III4. NATURALNESS IN THE WEINBERG-SALAM-GIM MODEL

The difficulties with the unnatural mass parameters only occur in theories with scalar fields. The only fundamental scalar

---

*) Conformal symmetry is violated at the quantum level.

field that occurs in the presently fashionable models is the Higgs field in the extended Weinberg-Salam model. The Higgs mass-squared, m_H^2, is up to a coefficient a fundamental parameter in the Lagrangian. It is small at energy scales $\mu \gg m_H$. Is there an approximate symmetry if $m_H \to o$? With some stretch of imagination we might consider a Goldstone-type symmetry:

$$\phi(x) \to \phi(x) + \text{const.} \tag{III8}$$

However we also had the local gauge transformations:

$$\phi(x) \to \Omega(x)\, \phi(x) \; . \tag{III9}$$

The transformations (III8) and (III9) only form a closed group if we also have invariance under

$$\phi(x) \to \phi(x) + C(x) \; . \tag{III10}$$

But then it becomes possible to transform ϕ away completely. The Higgs field would then become an unphysical field and that is not what we want. Alternatively, we could have that (III8) is an approximate symmetry only, and it is broken by all interactions that have to do with the symmetry (III9) which are the weak gauge field interactions. Their strength is $g^2/4\pi = \mathcal{O}(1/137)$. So at best we can have that the symmetry is broken by $\mathcal{O}(1/137)$ effects. Therefore

$$m_H^2/\mu^2 \gtrsim \mathcal{O}(1/137) \; .$$

Also the $\lambda\phi^4$ term in the Higgs field interactions breaks this symmetry. Therefore

$$m_H^2/\mu^2 \gtrsim \mathcal{O}(\lambda) \gtrsim \mathcal{O}(1/137) \; . \tag{III11}$$

Now

$$m_H^2 = \mathcal{O}(\lambda F_H^2) \; , \tag{III12}$$

where F_H is the vacuum expectation value of the Higgs field, known to be*)

$$F_H = (2G\sqrt{2})^{-1/2} = 174 \text{ GeV} \; . \tag{III13}$$

We now read off that

$$\mu \lesssim \mathcal{O}(F_H) = \mathcal{O}(174 \text{ GeV}) \; . \tag{III14}$$

---

*) Some numerical values given during the lecture were incorrect. I here give corrected values.

This means that at energy scales much beyond F_H our model becomes more and more unnatural. Actually, factors of π have been omitted. In practice one factor of 5 or 10 is still not totally unacceptable. Notice that the actual value of m_H dropped out, except that

$$m_H = \mathcal{O}\left(\frac{\sqrt{\lambda}}{g} M_W\right) \gtrsim \mathcal{O}(M_W) \ . \qquad \text{(III15)}$$

Values for m_H of just a few GeV are unnatural.

III5. EXTENDING NATURALNESS

Equation (III14) tells us that at energy scales much beyond 174 GeV the standard model becomes unnatural. As long as the Higgs field H remains a fundamental scalar nothing much can be done about that. We therefore conclude, with Dimopoulos and Susskind[2)] that the "observed" Higgs field must be composite. A non-trivial strongly interacting field theory must be operative at 1000 GeV or so. An obvious and indeed likely possibility is that the Higgs field H can be written as

$$H = Z\bar{\psi}\psi \ , \qquad \text{(III16)}$$

where Z is a renormalization factor and ψ is a new quark-like object, a fermion with a new color-like interaction [2)]. We will refer to the object as meta-quark having meta-color. The theory will have all features of QCD so that we can copy the nomenclature of QCD with the prefix "meta-". The Higgs field is a meta-meson.

It is now tempting to assume that the meta-quarks transform the same way under weak SU(2) x U(1) as ordinary quarks. Take a doublet with left-handed components forming one gauge doublet and right handed components forming two gauge singlets. The meta-quarks are massless. Suppose that the meta-chiral symmetry is broken spontaneously just as in ordinary QCD. What would happen?

What happens is in ordinary QCD well described by the Gell-Mann-Lévy sigma model. The lightest mesons form a quartet of real fields, ϕ_{ij}, transforming as a

$$2^{left} \otimes 2^{right}$$

representation of

$$SU(2)^{left} \otimes SU(2)^{right} .$$

Since the weak interaction only deals with $SU(2)^{left}$ this quartet can also be considered as one complex doublet representation of weak SU(2). In ordinary QCD we have

$$\phi_{ij} = \sigma\delta_{ij} + i\tau^a_{ij}\pi^a \quad , \tag{III17}$$

and

$$<\sigma>_{vacuum} = \frac{1}{\sqrt{2}} f_\pi = 91 \text{ MeV} . \tag{III18}$$

The complex doublet is then

$$\phi_i = \frac{1}{\sqrt{2}} \begin{pmatrix} \sigma + i\pi^3 \\ \pi^2 + i\pi^1 \end{pmatrix} , \tag{III19}$$

and

$$<\phi_i>_{vacuum} = \begin{pmatrix} 1 \\ 0 \end{pmatrix} \times 64 \text{ MeV} . \tag{III20}$$

We conclude that if we transplant this theory to the TeV range then we get a scalar doublet field with a non-vanishing vacuum expectation value for free. All we have to do now is to match the numbers. If we scale all QCD masses by a scaling factor κ then we match

$$F_H = 174 \text{ GeV} = \kappa \; 64 \text{ MeV} ;$$

$$\kappa = 2700 . \tag{III21}$$

Now the mesonic sector of QCD is usually assumed to be reproduced in the $1/N$ expansion [5] where N is the number of colors (in QCD we have N = 3). The 4-meson coupling constant goes like $1/N$. Then one would expect

$$f_\pi \propto \sqrt{N} . \tag{III22}$$

Therefore

$$\kappa = 2700 \sqrt{\frac{3}{N}} , \tag{III23}$$

if the metacolor group is SU(N).

Thus we obtain a model that reproduces the W-mass and predicts the Higgs mass. The Higgs is the meta-sigma particle. The ordinary sigma is a wide resonance at about 700 MeV[3], so that we predict

$$m_H = \kappa m_\sigma = 1900 \sqrt{\frac{3}{N}} \text{ GeV} , \tag{III24}$$

and it will be extremely difficult to detect among other strongly interacting objects.

III6. WHAT NEXT?

The model of the previous section is to our mind nearly inevitable, but there are problems. These have to do with the observed fermion masses. All leptons and quarks owe their masses to an interaction term of the form

$$g\bar{\psi}H\psi \, , \qquad \text{(III25)}$$

where g is a coupling constant, ψ is the lepton or quark and H is the Higgs field. With (III16) this becomes a four-fermion interaction, a fundamental interaction in the new theory. Because it is non-renormalizable further structure is needed. In ref. 2 the obvious choice is made: a new "meta-weak interaction" gauge theory enters with new super-heavy intermediate vector bosons. But since H is a scalar this boson must be in the crossed channel, a rather awkward situation. (See option a in Figure 1.) A simpler theory is that a new scalar particle is exchanged in the direct channel. (See option b in Figure 1.)

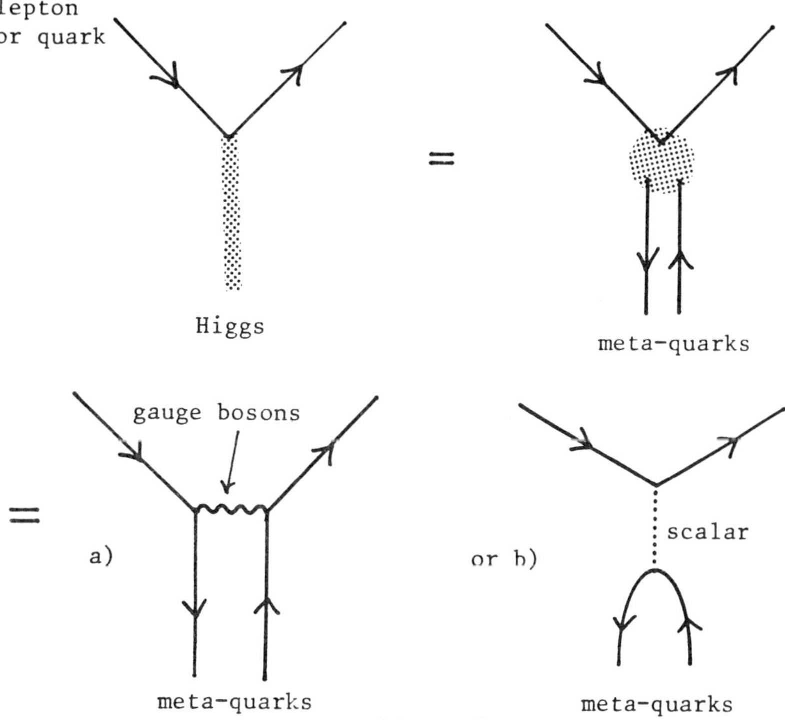

Figure 1.

Notice that in both cases new scalar fields are needed because in case a) something must cause the "spontaneous breakdown" of the new gauge symmetries. Therefore choice b) is simpler.
We removed a Higgs scalar and we get a scalar back. Does naturalness improve? The answer is yes. The coupling constant g in the interaction (III25) satisfies

$$g = g_1 g_2 / M_s^2 Z \ . \tag{III26}$$

Here g_1 and g_2 are the couplings at the new vertices, M_s is the new scalar's mass, and Z is from (III16) and is of order

$$Z \sim \frac{1}{\sqrt{\frac{N}{3}}\,(\kappa\, m_\rho)^2} = \frac{\sqrt{N/3}}{(1800\ \mathrm{GeV})^2} \ . \tag{III27}$$

Suppose that the heaviest lepton or quark is about 10 GeV. For that fermion the coupling constant g is

$$g = \frac{m_f}{F} \simeq 1/20 \ .$$

We get

$$g_1 g_2 \simeq \left(\frac{M_s}{1800\ \mathrm{GeV}}\right)^2 \sqrt{\frac{N}{3}} \cdot \frac{1}{20} \ .$$

Naturalness breaks down at

$$\mu = \mathcal{O}\left(\frac{M_s}{g_{1,2}}\right) = 8000 \ \sqrt[4]{\frac{3}{N}}\ \mathrm{GeV} \ ,$$

an improvement of about a factor 50 compared with the situation in sect. III4. Presumably we are again allowed to multiply by factors like 5 or 10, before getting into real trouble.

Before speculating on how to go on from here to improve naturalness still further we must assure ourselves that all other alleys are blind ones. An intriguing possibility is that the presently observed fermions are composite. We would get option c), Figure 2.

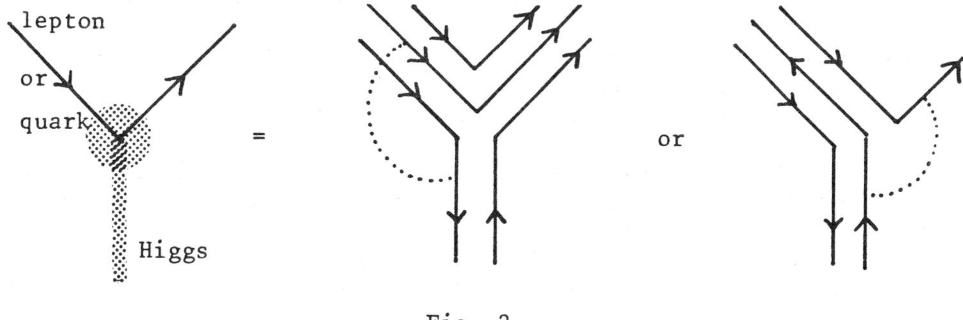

Fig. 2

The dotted line could be an ordinary weak interaction W or photon, that breaks an internal symmetry in the binding force for the new components. The new binding force could either act at the 1 TeV or at the 10-100 TeV range. It could either be an extension of meta-color or be a (color)" or paracolor force. Is such an idea viable?

Clearly, compared with the energy scale on which the binding forces take place, the composite fermions must be nearly massless. Again, this cannot be an accident. The chiral symmetry responsible for this must be present in the underlying theory. Apparently then, the underlying theory will possess a chiral symmetry which is not (or not completely) spontaneously broken, but reflected in the bound state spectrum in the Wigner mode: some massless chiral objects and parity doubled massive fermions. This possibility is most clearly described by the σ-model as a model for the lowest bound states occurring in ordinary quantum chromodynamics.

III7. THE σ MODEL

The fermion system in quantum chromodynamics shows an axial symmetry. To illuminate our problem let us consider the case of two flavors. The local color group is $SU(3)_c$. The subscript c here stands for color. The flavor symmetry group is $SU(2)_L \otimes SU(2)_R \otimes U(1)$ where the subscripts L and R stands for left and right and the group elements must be chosen to be space-time independent. We split the fermion fields ψ into left and right components:

$$\psi = \tfrac{1}{2}(1+\gamma_5)\psi_L + \tfrac{1}{2}(1-\gamma_5)\psi_R \ . \qquad \text{(III28)}$$

ψ_L transforms as a $3_c \otimes 2_L \otimes 1_R \otimes 2_{\mathcal{L}}$ (III29)

and ψ_R transforms as a $3_c \otimes 1_L \otimes 2_R \otimes \bar{2}_{\mathcal{L}}$ (III30)

where the indices refer to the various groups. $\mathcal{L}$ stands for the Lorentzgroup SO(3,1), locally equivalent to SL(2,c) which has two different complex doublet representations $2_{\mathcal{L}}$ and $\bar{2}_{\mathcal{L}}$ (corresponding to the transformation law for the neutrino and antineutrino, respectively). The fields ψ_L and ψ_R have the same charge under U(1), whereas axial U(1) group (under which they would have opposite charges) is absent because of instanton effects[4].

The effect of the color gauge fields is to bind these fermions into mesons and baryons all of which must be color singlets. It would be nice if one could describe these hadronic fields as representations of $SU(2)_L \otimes SU(2)_R \otimes U(1)$ and the Lorentz group, and then cast their mutual interactions in the form of an effective Lagrangian, invariant under the flavor symmetry group. In the case at hand this is possible and the resulting construction is a successful and one-time popular model for pions and nucleons: the σ model[3]. We have a nucleon doublet

$$N = \tfrac{1}{2}(1+\gamma_5)N_L + \tfrac{1}{2}(1-\gamma_5)N_R \ , \qquad \text{(III31)}$$

where

$$N_L \text{ transforms as a } 1_c \otimes 2_L \otimes 1_R \otimes 2_{\mathcal{L}} \ , \qquad \text{(III32a)}$$

$$\text{and } N_R \text{ transforms as a } 1_c \otimes 1_L \otimes 2_R \otimes \bar{2}_{\mathcal{L}} \ . \qquad \text{(III32b)}$$

Further we have a quartet of real scalar fields $(\sigma, \vec{\pi})$ which transform as a $1_c \otimes 2_L \otimes 2_R \otimes 1_{\mathcal{L}}$. The Lagrangian is

$$\mathcal{L} = -\bar{N}[\gamma\partial + g_o(\sigma + i\vec{\tau}.\vec{\pi}\gamma_5)]N - \tfrac{1}{2}(\partial\pi)^2 - \tfrac{1}{2}(\partial\sigma)^2 - V(\sigma^2+\vec{\pi}^2) \ . \qquad \text{(III33)}$$

Here V must be a rotationally invariant function.

Usually V is chosen such that its absolute minimum is away from the origin. Let V be minimal at $\sigma = v$ and $\vec{\pi} = o$. Here v is just a c-number. To obtain the physical particle spectrum we write

$$\sigma = v + s \qquad \text{(III34)}$$

and we find

$$\mathcal{L} = -\bar{N}(\gamma\partial + g_o v)N - \tfrac{1}{2}(\partial\vec{\pi})^2 - \tfrac{1}{2}(\partial s)^2 - 2v^2 V''(v^2)s^2 + \text{interaction terms} \ . \qquad \text{(III35)}$$

Clearly, in this case the nucleons acquire a mass term $m_s = g_o v$ and the s particle has a mass $m_s^2 = 4v^2V''(v^2)$, whereas the pion remains strictly massless. The entire mass of the pion must be due to effects that explicitly break $SU(2)_L \times SU(2)_R$, such as a small mass term $m_q\bar{\psi}\psi$ for the quarks (III28). We say that in this case the flavor group $SU(2)_L \otimes SU(2)_R$ is spontaneously broken into the isospin group SU(2).

Another possibility however, apparently not realised in ordinary quantum chromodynamics, would be that $SU(2)_L \otimes SU(2)_R$ is *not* spontaneously broken. We would read off from the Lagrangian (III33) that the nucleons N would form a massless doublet and that the four fields $(\sigma,\vec{\pi})$ could be heavy. The dynamics of other confining gauge theories could differ sufficiently from ordinary QCD so that, rather than a spontaneous symmetry breakdown, massless "baryons" develop. The principle question we will concentrate on is why do these massless baryons form the representation (III32), and how does this generalize to other systems. We would let future generations worry about the question where exactly the absolute minimum of the effective potential V will appear.

III8. INDICES

We now consider any color group G_c. The fundamental fermions in our system must be non-trivial representation of G_c and we assume "confinement" to occur: all physical particles are bound states that are singlets under G_c. Assume that the fermions are all massless (later mass terms can be considered as a perturbation). We will have automatically some global symmetry which we call the flavor group G_F. (We only consider exact flavor symmetries, not spoilt by instanton effects.) Assume that G_F is not spontaneously broken. Which and how many representations of G_F will occur in the massless fermion spectrum of the baryonic bound states? We must formulate the problem more precisely. The massless nucleons in (III33) being bound states, may have many massive excitations. However, massive Fermion fields cannot transform as a $2_\mathcal{L}$ under Lorentz transformations; they must go as a $2_\mathcal{L} \oplus \bar{2}_\mathcal{L}$. That is because a mass term being a Lorentz invariant product of two fields at one point only links $2_\mathcal{L}$ representations with $\bar{2}_\mathcal{L}$ representations. Consider a given representation r of G_F. Let p be the number of field multiplets transforming as $r \otimes 2_\mathcal{L}$ and q be the number of field multiplets $r \otimes \bar{2}_\mathcal{L}$. Mass terms that link the $2_\mathcal{L}$ with $\bar{2}_\mathcal{L}$ fields are completely invariant and in general to be expected in the effective Lagrangian. But the absolute value of

$$\ell = p - q \qquad \text{(III36)}$$

is the minimal number of surviving massless chiral field multiplets. We will call ℓ the index corresponding to the representation r of G_F. By definition this index must be a (positive or negative) integer. In the sigma model it is postulated that

$$\text{index } (2_L \otimes 1_R) = 1 \qquad \text{(III37)}$$

$$\text{index } (1_L \otimes 2_R) = -1$$

index (r) = o for all other representations r.

This tells us that if chiral symmetry is not broken spontaneously one massless nucleon doublet emerges. We wish to find out what massless fermionic bound states will come out in more general theories. Our problem is: how does (III37) generalize?

III9. ABSENCE OF MASSLESS BOUND STATES WITH SPIN 3/2 OR HIGHER

In the foregoing we only considered spin o and spin 1/2 bound states. Is it not possible that fundamentally massless bound states develop with higher spin? I believe to have strong arguments that this is indeed not possible. Let us consider the case of spin 3/2. Massive spin 3/2 fermions are described by a Lagrangian of the form

$$\mathcal{L} = \tfrac{1}{2}\bar{\psi}_\mu [\sigma_{\mu\nu}(\gamma\partial+m) + (\gamma\partial+m)\sigma_{\mu\nu}]\psi_\nu \ . \qquad \text{(III38)}$$

Just like spin-one particles, this has a gauge-invariance if $m \to o$:

$$\psi_\mu \to \psi_\mu + \partial_\mu \eta(x) \ , \qquad \text{(III39)}$$

where $\eta(x)$ is arbitrary. Indeed, massless spin 3/2 particles only occur in locally supersymmetric field theories. The field $\eta(x)$ is fundamentally unobservable.

Now in our model ψ_μ would be shorthand for some composite field: $\psi_\mu \to \psi\psi\psi$. However, then all components of this, including η, would be observables. If m = o we would be forced to add a gauge fixing term that would turn η into an unacceptable ghost particle*).

We believe, therefore, that unitarity and locality forbid the occurrence of massless bound states with spin 3/2. The case for higher spin will not be any better. And so we concentrate on a bound state spectrum of spin 1/2 particles only.

---

*) Note added: during the lectures it was suggested by one attendant to consider only gauge-invariant fields as $\Psi_{\mu\nu} = \partial_\mu\psi_\nu - \partial_\nu\psi_\mu$. However, such fields must satisfy constraints: $\partial[\alpha\Psi_{\mu\nu}] = o$. Composite field will never automatically satisfy such constraints.

III10. SPECTATOR GAUGE FIELDS AND -FERMIONS

So far, our model consisted of a strong interaction color gauge theory with gauge group G_C, coupled to chiral fermions in various representations r of G_c but of course in such a way that the anomalies cancel. The fermions are all massless and form multiplets of a global symmetry group, called G_F. For QCD this would be the flavor group. In the metacolor theory G_F would include all other fermion symmetries besides metacolor.

In order to study the mathematical problem raised above we will add another gauge connection field that turns G_F into a local symmetry group. The associated coupling constants may all be arbitrarily small, so that the dynamics of the strong color gauge interactions is not much affected. In particular the massless bound state spectrum should not change. One may either think of this new gauge field as a completely quantized field or simply as an artificial background field with possibly non-trivial topology. We will study the behavior of our system in the presence of this "spectator gauge field". As stated, its gauge group is G_F.

Note however, that some flavor transformations could be associated with anomalies. There are two types of anomalies:

i) those associated with $G_c \times G_F$, only occurring where the color field has a winding number. Only U(1) invariant subgroups of G_F contribute here. They simply correspond to small explicit violations of the G_F symmetry. From now on we will take as G_F only the anomaly-free part. Thus, for QCD with N flavors, G_F is not $U(N) \times U(N)$ but

$$G_F = SU(N) \otimes SU(N) \otimes U(1) .$$

ii) those associated with G_F alone. They only occur if the spectator gauge field is quantized. To remedy these we simply add "spectator fermions" coupled to G_F alone. Again, since these interactions are weak they should not influence the bound state spectrum.

Here, the spectator gauge fields and fermions are introduced as mathematical tools only. It just happens to be that they really do occur in Nature, for instance the weak and electromagnetic $SU(2) \times U(1)$ gauge fields coupled to quarks in QCD. The leptons then play the role of spectator fermions.

III11. ANOMALY CANCELLATION FOR THE BOUND STATE SPECTRUM

Let us now resume the particle content of our theory. At small distances we have a gauge group $G_C \otimes G_F$ with chiral fermions in several representations of this group. Those fermions which are trivial under G_C are only coupled weakly and are called "spectator fermions". All anomalies cancel, by construction.

At low energies, much lower than the mass scale where color binding occurs, we see only the G_F gauge group with its gauge fields. Coupled to these gauge fields are the massless bound states, forming new representations r of G_F, with either left- or right handed chirality. The numbers of left minus right handed fermion fields in the representations r are given by the as yet unknown indices $\ell(r)$. And finally we have the spectator fermions which are unchanged.

We now expect these very light objects to be described by a new local field theory, that is, a theory local with respect to the large distance scale that we now use. The central theme of our reasoning is now that this new theory must again be anomaly free. We simply cannot allow the contradictions that would arise if this were not so. Nature must arrange its new particle spectrum in such a way that unitarity is obeyed, and because of the large distance scale used the effective interactions are either vanishingly small or renormalizable. The requirement of anomaly cancellation in the new particle spectrum gives us equations for the indices $\ell(r)$, as we will see.

The reason why these equations are sometimes difficult or impossible to solve is that the new representations r must be different from the old ones; if $G_C = SU(N)$ then r must also be faithful representations of $G_F/Z(N)$. For instance in QCD we only allow for octet or decuplet representations of $(SU(3))_{flavor}$, whereas the original quarks were triplets.

However, the anomaly cancellation requirement, restrictive as it may be, does not fix the values of $\ell(r)$ completely. We must look for additional limitations.

III12 APPELQUIST-CARAZZONE DECOUPLING AND N-INDEPENDENCE

A further limitation is found by the following argument. Suppose we add a mass term for one of the colored fermions.

$$\Delta\mathcal{L} = m\,\bar{\psi}_{1L}\,\psi_{1R} + \text{h.c.}$$

Clearly this links one of the left handed fermions with one of the right handed ones and thus reduces the flavor group G_F into $G_F' \subset G_F$. Now let us gradually vary m from o to infinity. A famous theorem [5] tells us that in the limit $m \to \infty$ all effects due to this massive quark disappear. All bound states containing this quark should also disappear which they can only do by becoming very heavy. And they can only become heavy if they form representations r' of G_F' with total index $\ell'(r') = o$. Each representation r of G_F forms an array of representations r' of G'_F. Therefore

$$\ell'(r') = \sum_{r \text{ with } r' \subset r} \ell(r) \ . \qquad \text{(III40)}$$

Apparently this expression must vanish.

Thus we found another requirement for the indices $\ell(r)$. The indices will be nearly but not quite uniquely determined now. Calculations show that this second requirement makes our indices $\ell(r)$ practically independent of the dimensions n_i of G_F. For instance, if G_c = SU(3) and if we have left- and righthanded quarks forming triplets and sextets then

$$G_F = SU(n_1)_L \otimes SU(n_2)_R \otimes SU(n_3)_L \otimes SU(n_4)_R \otimes U(1)^3 \qquad \text{(III41)}$$

where $n_{1,2}$ refer to the triplets and $n_{3,4}$ to the sextets. G_c is anomaly-free if

$$n_1 - n_2 + 7(n_3 - n_4) = o \ . \qquad \text{(III42)}$$

Here we have three independent numbers n_i.
If we write the representations r as Young tableaus then $\ell(r)$ could still depend explicitly on n_i.

However, suppose that someone would start as approximation of Bethe-Salpeter type to discover the zero mass bound state spectrum. He would study diagrams such as Fig. 3

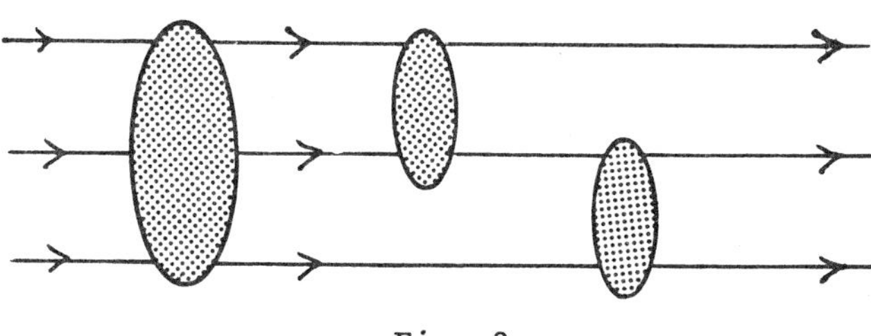

Fig. 3

The resulting indices $\ell(r)$ would follow from topological properties of the interactions represented by the blobs. It is unlikely that this topology would be seriously effected by details such as the contributions of diagrams containing additional closed fermion loops. However, that is the only way in which explicit n-dependence enters. It is therefore natural to assume $\ell(r)$ to be n-independent. This latter assumption fixes $\ell(r)$ completely. What is the result of these calculations?

III13. CALCULATIONS

Let G be any (reducible or irreducible) gauge group. Let chiral fermions in a representation r be coupled to the gauge fields by the covariant derivative

$$D_\mu = \partial_\mu + i\,\lambda^a(r)\,A^a_\mu \;, \qquad \text{(III43)}$$

where A^a_μ are the gauge fields and $\lambda^a(r)$ a set of matrices depending on the representation r. Let the left-handed fermions be in the representations r_L and the right-handed ones in r_R. Then the anomalies cancel if

$$\sum_L \mathrm{Tr}\{\lambda^a(r_L),\, \lambda^b(r_L)\}\,\lambda^c(r_L) =$$

$$\sum_R \mathrm{Tr}\{\lambda^a(r_R),\, \lambda^b(r_R)\}\,\lambda^c(r_R)\;. \qquad \text{(III44)}$$

The object $d^{abc}(r) = \mathrm{Tr}\{\lambda^a(r),\, \lambda^b(r)\}\,\lambda^c(r)$ can be computed for any r. In table 1 we give some examples. The fundamental representation r_o is represented by a Young tableau: □ . Let it have n components. We take the case that $\mathrm{Tr}\,\lambda(r_o) = o$. Write

$$\mathrm{Tr}\,I(r_o) = n\;, \qquad \mathrm{Tr}\,I(r) = N(r)\;,$$

$$\mathrm{Tr}\,\lambda(r) = o\;,$$

$$\mathrm{Tr}\,\lambda^a(r)\,\lambda^b(r) = C(r)\,\mathrm{Tr}\,\lambda^a(r_o)\,\lambda^b(r_o)\;,$$

$$d^{abc}(r) = K(r)\,d^{abc}(r_o)\;. \qquad \text{(III45)}$$

We read off C and K from table 1.
Now III44 must hold both in the high energy region and in the low energy region. The contribution of the spectator fermions in both regions is the same. Thus we get for the bound states

$$\left(\sum_L - \sum_R\right) d^{abc}(r) = n_c\left(d^{abc}(r_{oL}) - d^{abc}(r_{oR})\right) \qquad \text{(III46)}$$

where a,b,c are indices of G_F and r_o is the fundamental representation of G_F. We have the factor n_c written explicitly, being the number of color components.

Let us now consider the case $G_c = SU(3)$;
$G_F = SU_L(n) \otimes SU_R(n) \otimes U(1)$. We have n "quarks" in the fundamental representations. The representations r of the bound states must be in $G_F/Z(3)$. They are assumed to be built from three quarks, but we are free to choose their chirality. The expected representations

Table 1

r	N(r)	C(r)	K(r)
□	n	1	1
□	n	1	-1
□□ / □□ (column)	$\frac{n(n\pm 1)}{2}$	$n\pm 2$	$n\pm 4$
□□□ / □□□ (column)	$\frac{n(n\pm 1)(n\pm 2)}{6}$	$\frac{(n\pm 2)(n\pm 3)}{2}$	$\frac{(n\pm 3)(n\pm 6)}{2}$
□□ over □	$\frac{n(n^2-1)}{3}$	n^2-3	n^2-9
$A \otimes B$	N(A)N(B)	C(A)N(B) + C(B)N(A)	K(A)N(B) + K(B)N(A)

are given in table 2, where also their indices are defined. Because of left-right symmetry these numbers change sign under interchange of left ↔ right.

Table 2

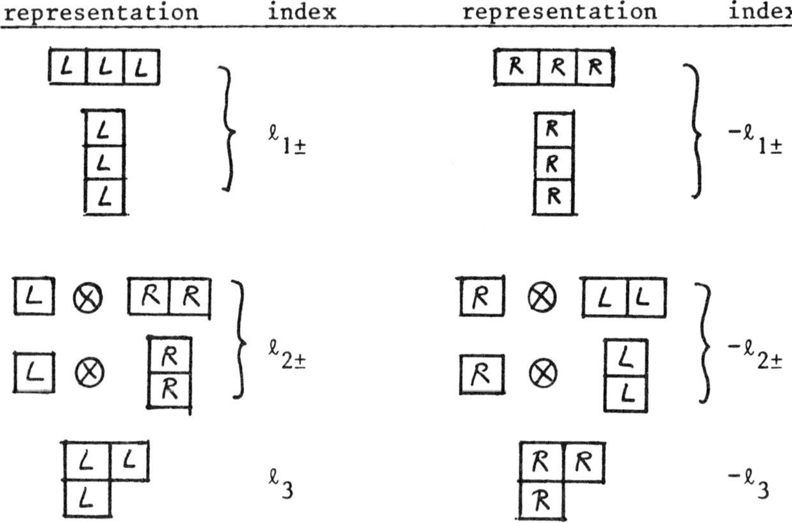

For the time being we assume no other representations. In eq. III46 we may either choose a, b and c all to be $SU(n)_L$ indices, or choose a and b to be $SU(n)_L$ indices and c the U(1) index. We get two independent equations:

$$\sum_{\pm} \tfrac{1}{2}(n\pm3)(n\pm6)\ell_{1\pm} - \sum_{\pm} \tfrac{1}{2}n(n\pm7)\ell_{2\pm} + (n^2-9)\ell_3 = 3, \text{ if } n > 2 ,$$

and

$$\sum_{\pm} \tfrac{1}{2}(n\pm2)(n\pm3)\ell_{1\pm} - \sum_{\pm} \tfrac{1}{2}n(n\pm3)\ell_{2\pm} + (n^2-3)\ell_3 = 1, \text{ if } n > 1 . \qquad \text{(III47)}$$

The Appelquist-Carazzone decoupling requirement, eq. (III40), gives us in addition two other equations:

$$\ell_{1+} - \ell_{2+} + \ell_3 = 0 ,$$

$$\ell_{1-} - \ell_{2-} + \ell_3 = 0 , \text{ both if } n > 2 . \qquad \text{(III48)}$$

For n > 2 the general solution is

$$\ell_{1+} = \ell_{1-} = \ell ,$$
$$\ell_{2+} = \ell_{2-} = 3\ell - \frac{1}{3} ,$$
$$\ell_3 = 2\ell - \frac{1}{3} . \qquad \text{(III49)}$$

Here ℓ is still arbitrary. Clearly this result is unacceptable. We cannot allow any of the indices ℓ to be non-integer. Only for the case n = 2 (QCD with just two flavors) there is another solution. In that case ℓ_2 and ℓ_3 describe the same representation, and ℓ_{1-} an empty representation. We get

$$\ell_{2-} + \ell_3 = k = 1 - 10\ \ell_{1+} + 5\ \ell_{2+} . \qquad \text{(III50)}$$

According to the σ-model, $\ell_{1+} = \ell_{2+} = o$; $k = 1$. The σ-model is therefore a correct solution to our equations.

In the previous section we promised to determine the indices completely. This is done by imposing n-independence for the more general case including also other color representations such as sextets besides triplets. The resulting equations are not very illuminating, with rather ugly coefficients. One finds that in general no solution exists except when one assumes that all mixed representations have vanishing indices. With mixed representations we mean a product of two or more non-trivial representations of two or more non-Abelian invariant subgroups of G_F. If now we assume n-independence this must also hold if the number of sextets is zero. So ℓ_{2+} and ℓ_{2-} must vanish. We get

$$\ell_{1+} = \ell_{1-} = 1/9 ,$$
$$\ell_3 = -1/9 . \qquad \text{(III51)}$$

If all quarks were sextets, not triplets, we would get

$$\ell_{1+} = \ell_{1-} = 2/9 ,$$
$$\ell_3 = -2/9 . \qquad \text{(III52)}$$

In the case $G_C = SU(5)$ the indices were also found. See table 3.

Table 3
indices for G_c = SU(5)

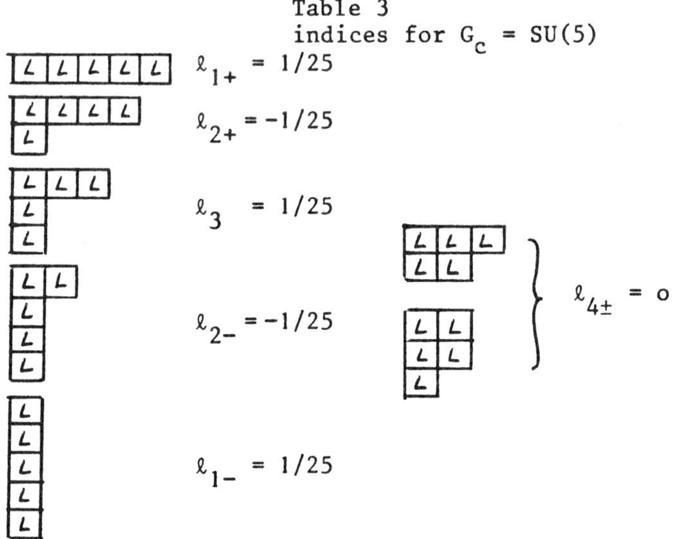

This clearly suggests a general tendency for SU(N) color groups to produce indices $\pm 1/N^2$ or o.

III14. CONCLUSIONS

Our result that the indices we searched for are fractional is clearly absurd. We nevertheless pursued this calculation in order to exhibit the general philosophy of this approach and to find out what a possible cure might be. Our starting point was that chiral symmetry is not broken spontaneously. Most likely this is untenable, as several authors have argued[6)]. We find that explicit chiral symmetry in QCD leads to trouble in particular if the number of flavors is more than two. A daring conjecture is then that in QCD the strange quark, being rather light, is responsible for the spontaneous breakdown of chiral symmetry.
An interesting possibility is that in some generalized versions of QCD chiral symmetry is broken only partly, leaving a few massless chiral bound states. Indeed there are examples of models where our philosophy would then give integer indices, but since we must drop the requirement of n-dependence our result was not unique and it was always ugly. No such model seems to reproduce anything resembling the observed quark-lepton spectrum.

Finally there is the remote possibility that the paradoxes associated with higher spin massless bound states can be resolved. Perhaps the Δ(1236) plays a more subtle role in the σ-model than assumed so far (we took it to be a parity doublet).

We conclude that we are unable to construct a bound state theory for the presently fundamental fermions along the lines suggested above.

We thank R. van Damme for a calculation yielding the indices in the case G_c = SU(5).

REFERENCES

1. P.A.M. Dirac, Nature 139 (1937) 323, Proc. Roy. Soc. A165 (1938) 199, and in: Current Trends in the Theory of Fields, (Tallahassee 1978) AIP Conf. Proc. No 48, Particles and Fields Subseries No 15, ed. by Lannuti and Williams, p. 169.
2. S. Dimopoulos and L. Susskind, Nucl. Phys. B155 (1979) 237.
3. M. Gell-Mann and M. Lévy, Nuovo Cim. 16 (1960) 705.
   B.W. Lee, Chiral Dynamics, Gordon and Breach, New York, London, Paris 1972.
4. G. 't Hooft, Phys. Rev. Lett. 37 (1976) 8; Phys. Rev. D14 (1976) 3432.
   S. Coleman, "The Uses of Instantons", Erice Lectures 1977.
   R. Jackiw and C. Rebbi, Phys. Rev. Lett. 37 (1976) 172.
   C. Callan, R. Dashen and D. Gross, Phys. Lett. 63B (1976) 334.
5. G. 't Hooft, Nucl. Phys. B72 (1974) 461.
6. T. Appelquist and J. Carazzone, Phys. Rev. D11 (1975) 2856.
7. A. Casher, Chiral Symmetry Breaking in Quark Confining Theories, Tel Aviv preprint TAUP 734/79 (1979).

Volume 99B, number 4 PHYSICS LETTERS 26 February 1981

A QUANTUM STRUCTUREDYNAMIC MODEL OF QUARKS, LEPTONS, WEAK VECTOR BOSONS AND HIGGS MESONS ☆

O.W. GREENBERG and Joseph SUCHER

Center for Theoretical Physics, Department of Physics and Astronomy, University of Maryland, College Park, MD 20742, USA

Received 14 August 1980
Revised manuscript received 13 November 1980

We propose a model in which quarks, leptons and weak vector bosons are two-body composites and Higgs mesons are two- or many-body composites confined by an SU(N) local gauge "quantum structuredynamics" interaction, quark–lepton parallelism holds, and the strong and weak interactions are on a similar footing -- both are residual effects of a flavor-independent local gauge interaction.

Several authors have suggested that composite structure of quarks and leptons is necessary to understand quark–lepton parallelism and the repetition of generations. We propose a model in which weak vector bosons and Higgs mesons are composite as well as quarks and leptons [‡1].

The fundamental fields in our model are a spin-1/2 flavor doublet $F = (F_u, F_d)$ of flavons together with two flavor singlets, a spin-0 $SU(3)_c$ antitriplet chromon C and a spin-0 "leptoscalar" S. The electric charges (in terms of $|e|$) are 2/3, −1/3, for F_u, F_d, 0 for C and 2/3 for S [‡2]. We assume that F, C, and S belong to the fundamental **N** representation of a confining local gauge group $SU(N)_S$ which is responsible for the existence and structure of quarks, leptons and weak bosons. The lagrangian for the "quantum structuredynamics" [‡3] (QSD) of the $SU(N)_S$ local gauge-invariant interaction is

$$\mathcal{L} = -\tfrac{1}{4} G^a_{\mu\nu} G^{a\mu\nu} - \tfrac{1}{4} F^\alpha_{\mu\nu} F^{\alpha\mu\nu} + \bar{F}_i (i \not\partial + g_S \not{\mathcal{G}}) F_i + |(\partial_\mu - i g_S \mathcal{G}_\mu - i g_c \mathcal{F}_\mu) C|^2 + |(\partial_\mu - i g_S \mathcal{G}_\mu) S|^2$$
$$- \tfrac{1}{2} m_c^2 C^\dagger_\alpha C^\alpha - \tfrac{1}{2} m_S^2 S^\dagger S - \lambda_c (C^\dagger_\alpha C^\alpha)^2 - \lambda_{cS} C^\dagger_\alpha C^\alpha S^\dagger S - \lambda_S (S^\dagger S)^2 \,, \tag{1}$$

where the $SU(N)_S$ and $SU(3)_c$ gauge fields and potentials are related by

$$G^a_{\mu\nu} = \partial_\mu G^a_\nu - \partial_\nu G^a_\mu + g_S f^{abc} G^b_\mu G^c_\nu \,, \quad a, b, c = 1, 2, \ldots, N \,, \tag{2}$$

$$\mathcal{G}_\mu = \tfrac{1}{2} \lambda^a G^a_\mu \,, \quad \text{where } \lambda^a\text{'s are SU}(N)\ \lambda\text{'s} \,, \tag{3}$$

$$F^\alpha_{\mu\nu} = \partial_\mu F^\alpha_\nu - \partial_\nu F^\alpha_\mu + g_c f^{\alpha\beta\gamma} F^\beta_\mu F^\gamma_\nu \,, \quad \alpha, \beta, \gamma = 1, 2, 3 \,, \tag{4}$$

$$\mathcal{F}_\mu = \tfrac{1}{2} \lambda^\alpha F^\alpha_\mu \,, \quad \text{where } \lambda^\alpha\text{'s are SU}(3)\ \lambda\text{'s} \,, \tag{5}$$

☆ A preliminary report of this work was given at the 20th International Conference on High Energy Physics, at Madison, WI (July 1980), to be published in the Proceedings of the Conference.

‡1 We give references to composite models of quarks and/or leptons in ref. [1].

‡2 A more general possible charge pattern is $q + 1$, q, $q + 1/3$, $q + 1$ for F_u, F_d, C, S, respectively. The asymptotic value of the ratio $\sigma(e^+e^- \to \text{hadrons})/\sigma(e^+e^- \to \mu\bar{\mu})$ is $(33q^2 + 30q + 13)/(27q^2 + 30q + 15)$.

‡3 We use "quantum structuredynamics" to characterize the dynamics of the internal structure of quarks, leptons, etc. which is associated with confinement by an SU(N) local gauge interaction.

Volume 99B, number 4 PHYSICS LETTERS 26 February 1981

and $F_i = (F_u, F_d)$. We expect QSD to give permanent confinement of the F's, C's, and S's into $SU(N)_S$ singlets. We conjecture that the size r_0 of the composites will be much smaller than the typical size $r_h = \hbar/m_\rho c$ of hadrons or the Compton wavelength $r_\ell = \hbar/m_\ell c$ of leptons provided that in $SU(N)_S \times SU(3)_c$, N is greater than three [‡4]. We assign the first generation of quarks u, d and leptons ν_e, e to the lowest $J = 1/2$ states of $(F_u\bar{C})$, $(F_d\bar{C})$ and $(F_u\bar{S})$, $(F_d\bar{S})$, respectively. We assign higher generations of quarks and leptons to the $J = 1/2$ excited states of these systems. Higher states will be narrow for masses below the W mass. In addition to the local $\mathcal{H} = SU(3)_c \times SU(N)_S$ symmetry, $\mathcal{L}$ has global $\mathcal{G} = SU(2)_L \times SU(2)_R \times U(1)$ flavor symmetry which gives a chiral classification of the $\bar{F}F$ bound states. There will be two-body scalar composites $\bar{F}_L F_R = \psi$ which we identify as Higgs fields of the effective low-energy theory, as well as vector composites $\bar{F}_L\gamma_\mu\tau^i F_L = W^i_{L\mu}$ and $\bar{F}_R\gamma_\mu\tau^i F_R = W^i_{R\mu}$, $i = 1, 2\ 3$, and $\bar{F}\gamma_\mu F$ which we identify, before spontaneous symmetry breaking, as massless gauge bosons of the effective low-energy theory which have $V \mp A$ and pure V couplings, respectively. There will also be N-body scalar composites which we identify as additional Higgs fields of the low-energy theory; for example for $N = 4$,

$$\phi_L = [(F^T_{La}C^{-1}\tau^i\gamma_\mu F_{Lb})(F^T_{Lc}C^{-1}\tau^j\gamma^\mu F_{Ld}) - \tfrac{1}{3}\delta^{ij}(F^T_{La}C^{-1}\tau^k\gamma^\mu F_{Lb})(F^T_{Lc}C^{-1}\tau^k\gamma^\mu F_{Ld})]\,\epsilon^{abcd}\,, \quad C = i\gamma^2\gamma^0\,, \tag{6}$$

with an analogous expression for ϕ_R; both ϕ's are flavor–spin 2 composite scalar fields [‡5]. We accept the conjecture [‡6] that, when the size r_0 of the composites is much less than their Compton wavelength $\hbar/mc$, the lagrangian $\mathcal{L}$[eq. (1)] of the fundamental fields will give rise to a renormalizable local effective low energy lagrangian $\mathcal{L}_{eff}$ for the composite fields, which include the different generations of quarks and leptons, the $SU(2)_L \times SU(2)_R \times U(1)$ vector bosons, and the Higgs mesons ψ, ϕ_L and ϕ_R. Renormalizability implies that $\mathcal{L}_{eff}$ is a spontaneously broken local, $\mathcal{G}$ gauge-invariant lagrangian with left–right symmetry, whose reduction to the standard $SU(2)_L \times U(1)$ theory has been discussed in ref. [4].

All the two-body $SU(N)_S$ singlets in our model, except $\bar{C}C$, $\bar{S}S$, $\bar{S}C$, and $\bar{C}S$, which we assume lie high in mass, correspond to quarks, leptons, weak vector bosons, or Higgs mesons. We can either keep the photon and/or the $SU(3)_c$- octet of gluons as elementary gauge bosons, consistent with the idea that exact local symmetries are associated with elementary fields, or construct the photon as $[2\bar{F}_u\gamma_\mu F_u - \bar{F}_d\gamma_\mu F_d + 2iS^\dagger\overleftrightarrow{\partial}_\mu S]/3$ and/or the color gluons as $C^\dagger\overleftrightarrow{\partial}_\mu\lambda^\alpha C$ composites. In this latter case, the QSD annihilation graphs would have to give mass to the SU(3)-singlet gluon, leaving the color octet massless. If the photon is elementary, electromagnetism can break global flavor symmetry in the usual way.

‡4 We assume $SU(3)_c$ and $SU(N)_S$ have the same strength at a large mass scale [3]. Various extended color schemes have been suggested to achieve small ratios of size (or, equivalently, large mass scale ratios) in ref. [2]. Assuming that F_u and F_d are light, and S and C are very heavy, and only the gauge multiplet of the unifying group $\mathcal{U}$ contains both light and heavy particles, the renormalization group equations lead to $N \approx 3 + \ln(M_N^2/M_3^2)/\ln(M_u^2/M_N^2)$, when M_3 and M_N, are the masses at which $SU(3)_c$ and $SU(N)_S$ become confining (i.e., the relevant $\alpha \approx 1$) and M_u is the mass at which the coupling constants coalesce. For example, the minimum value $N = 4$ requires $M_u = 10^{10}$ GeV if we choose $M_N = 10^5$ GeV and $M_3 = 1$ GeV. Larger values of N allow smaller M_u.

‡5 The form (6) is an $SU(N)_S$ singlet, a Lorentz scalar, and a flavor non-singlet. Simpler expressions constructed from the F_L's fail to have these three properties.

‡6 M. Veltman, quoted in Ellis et al. [1]. Related issues have been discussed by 't Hooft, Dimopoulos et al. [1], and Raby et al. [2]. We thank J.C. Pati for emphasizing the relevance of these ideas.

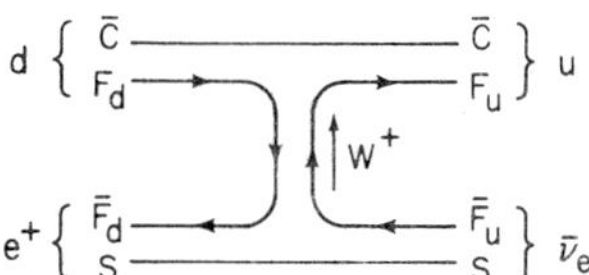

Fig. 1. The reaction $de^+ \to u\bar{\nu}_e$ (a crossed reaction of $d \to ue^-\bar{\nu}_e$) mediated by composite W^+ exchange.

Volume 99B, number 4 PHYSICS LETTERS 26 February 1981

A major feature of our point of view is that the weak interactions of quarks and leptons should follow as residual effects of flavor-independent QSD just as strong interactions are considered to follow as residual effects of flavor-independent QCD. For an example, see fig. 1. Thus the parallelism in which both the strong and weak interactions are associated with flavor-independent local gauge symmetries supplements the quark–lepton parallelism (in the counting of states) which follows from the introduction of the flavons [‡7].

Whether the scheme presented above is really capable of leading to the observed mass spectrum of the leptons and the "observed" mass spectrum of the quarks and weak bosons is an open question [‡8]. The virtue of a composite model such as ours based on a confining gauge theory is that, in principle, the spectrum and interactions of the composites are determined by the gauge theory and if the conjecture of footnote 6 holds the effective theory of the composites at low energy will be a local renormalizable gauge theory. The drawback is that we are no better able to do non-perturbative calculations of the properties of the composites in our model than we can for hadrons in QCD, and we do not know how the conjecture of footnote 6 can be derived as a property of the vertex functions of the effective low-energy theory, and if α_s, α_{em}, and G_F will have their observed magnitudes. We can discuss our composites in a qualitative way, give conditions which, if met, would insure that our model agrees with known weak interaction phenomenology, and check that no contradictions with experiment arise at a kinematic level.

We use the basis of flavor-singlet and flavor-triplet $(\bar{F}_L F)$, $(\bar{F}_R F_R)$, $(\bar{F}_L F_R) \pm (\bar{F}_R F_L)$ states in our heuristic discussion of weak bosons. Using the lore that states with additional ∂^μ factors lie higher than those with 1 or γ^μ, we expect that the lowest states, in the point limit, will be 7 $J = 1$ states of the form $(\bar{F}_L \gamma^\mu F_L)$, $(\bar{F}_R \gamma^\mu F_R)$, and $(\bar{F}\gamma^\mu F)$, and 8 $J = 0$ states of the form $(\bar{F}_L 1 F_R)$ and $(\bar{F}_R 1 F_L)$. To reproduce charged current phenomenology, the $J = 0$ states together with the two charged $(\bar{F}_R \gamma^\mu F_R)$ states must lie much higher than the charged $(\bar{F}_L \gamma^\mu F_L)$ states, which we identify with the $W^\pm$. To reproduce neutral current phenomenology, we need to use at least the neutral member of the left-handed triplet and the singlet $(\bar{F}\gamma^\mu F)$ states. These two states and the neutral member of the right-handed triplet mix because of flavor-symmetry breaking. The photon also mixes with these states, and if the photon mixing is sufficiently large [5], only the neutral $(\bar{F}_L \gamma^\mu F_L)$ state is needed. Preliminary investigation of the triangle graph for the vertex for lepton (quark) → lepton (quark) + weak boson in terms of the composite vertex functions and the F and S (C) propagators shows that the charged $(\bar{F}_{L,R} \gamma^\mu F_{L,R})$ weak boson states will have $V \mp A$ weak interactions, provided the F propagators and lepton (quark) vertex functions, which have the form $a + b\not{k}$, are dominated by the $\not{k}$ terms. Analogous conditions using the elementary electromagnetic field instead of a composite weak boson give a phenomenological γ^μ electromagnetic vertex for quarks and leptons [‡9]. The observed suppression of $\mu \to e\gamma$ requires a size $r_0 \lesssim (10^8\ \text{TeV})^{-1}$ if we estimate the matrix element as $em_\mu r_0$, and $r_0 \lesssim (10^2\ \text{TeV})^{-1}$ if we estimate it as $e(m_\mu r_0)^2$. Both of these distances are smaller than the size $r_0' \lesssim (5\ \text{TeV})^{-1}$ required to satisfy the experimental bound on $g - 2$.

To discuss the Cabibbo matrix and the GIM mechanism, we introduce phenomenological $SU(2)_f$ flavor currents, $j^i_{L\mu} = \frac{1}{2}\bar{F}_L \tau^i \gamma_\mu F_L$, and their associated charges which satisfy the $SU(2)_L$ algebra, $[Q^+_L, Q^-_L] = 2Q^3_L$. The charge-raising zero momentum-transfer Cabibbo matrix is $C_{ij} = \langle u_i | Q^+_L | d_j \rangle$, where $i, j = 1, 2, 3, \ldots$ label the generations. Provided single-quark intermediate states dominate, the $SU(2)_L$ algebra leads to the sum rule

‡7 Alternatively, the local gauge symmetry for the weak and electromagnetic interactions could be introduced at the level of $\mathcal{L}$ by gauging the flavor degree of freedom and introducing fundamental Higgs mesons. Then this parallelism would not hold, unless the fundamental gauge bosons were to acquire much larger masses (via spontaneous symmetry breaking) than the composite vector bosons.

‡8 The problem of the mass spectrum includes the following issues: why are quarks, leptons, weak vector bosons, etc., so light in mass compared to the large mass scales of the grand unified theories and of the inverse size r_0^{-1} of the composites; why are the mass spacings between quarks and leptons so small compared to r_0^{-1}; why are weak vector bosons much more massive than quarks and leptons; why do the quark and lepton mass spacings increase rapidly with generations; why do all quarks and leptons have spin 1/2 (assuming the spin of τ, which is not yet established, is 1/2); why do all neutrinos have very small or vanishing masses; why are not electromagnetic mass differences very large, of order e^2/r_0, if the photon is elementary? A problem specific to our model is: why are $\bar{S}C$, $\bar{C}S$, etc. very massive?

‡9 Electromagnetic properties of composite models have been discussed recently in ref. [6].

Volume 99B, number 4 PHYSICS LETTERS 26 February 1981

$$\delta_{ij} = \langle u_i | [Q_L^+, Q_L^-] | u_j \rangle \approx \sum_k C_{ik} C_{jk}^* \,, \tag{7}$$

which states that C is approximately unitary. If, in addition, the $Q_L^\pm$ operators change the generation i by at most one and the matrix elements $\langle u_i | Q_L^+ | d_j \rangle$ are real, we recover the GIM structure of weak interactions. The elements of the full Cabibbo matrix, which are determined by the overlap of the quark vertex functions for the different generations, are analogous to quasi-elastic form factors. We expect the diagonal elements to decrease with increasing momentum transfer [‡10].

There is no reason for neutrinos to be massless in our model [‡11].

We comment briefly on possible particle assignments for the choice of SU(N + 3) as the unifying group $\mathcal{U}$. We decompose representations of SU(N + 3) as (dim $SU(N)_S$, dim $SU(3)_c$) under the subgroup $SU(N)_S \times SU(3)_c$. The antiflavons (rather than the flavons) $\bar{F}_u$ and $\bar{F}_d$ could each be an (N, 1) part of an N + 3 fundamental representation, accompanied in each case by a (1, 3) color antiflavon. Similarly, the antileptoscalar $\bar{S}$ could be an (N, 1) part of a fundamental representation, accompanied by a (1,3) color antileptoscalar. The antichromon could be an (N, 3) part of the rank-two antisymmetric representatoion $(N + 3)(N + 2)/2$, accompanied by an $(N(N - 1)/2, 1)$ and a $(1, 3^*)$. The $SU(N)_S$ and $SU(3)_c$ gluons could appear as $(N^2 - 1, 1)$ and (1,8), respectively, in the adjoint representation of SU(N + 3), together with an $(N, 3^*)$, and $(N^*, 3)$ and a (1,1). (Since both spin-0 and spin-1/2 fields are present in our model, an alternative approach to unification might be via supersymmetry.)

Our model differs from most substructure models in making quarks and leptons out of a spin-1/2 and a spin-0 constituent, rather than using only spin-1/2 constituents. An advantage of using spin-0 constituents is that we avoid the $J = 3/2$ states which are likely to occur due to spin re-orientations, at comparable mass as $J = 1/2$ states if quarks and leptons are made of three or more spin-1/2 particles. (If, as conjectured, footnote 6, the effective low-energy theory is necessarily renormalizable and therefore free of fermions with $J > 1/2$, higher-J would be absent in any case.) However, even with a spin-0 constituent there remains the problem of avoiding low-lying higher-J states arising from orbital excitations. As indicated by Nowak et al. [1], this can be avoided if the effective two-body potential is sufficiently short-range and singular at the origin.

The applicability of non-relativistic potential models to the structure of quarks and leptons is problematic, because the bound states are likely to be relativistic. In particular, we see no reason for our QSD model to lead to a non-relativistic picture. Nonetheless, as background information we comment briefly on such a picture. For the explicit example of a non-relativistic particle bound in a potential $\sim r^\alpha$, in the semiclassical approximation [7] for the E_{nl}, where n is the principal quantum number and l is the orbital angular momentum, the requirement $E_{11} > E_{30}$, i.e., the first orbital excitation lies above the first three quark or lepton generations, requires $\alpha < -3/2$. To have n generations below the first orbital excitation requires $\alpha < -(2n - 3)/(n - 1)$. A simple way to see that the elevation of orbital excitations relative to $l = 0$ states imposes special constraints on the binding interaction is to note that the uncertainty principle applied to the radial coordinate, $rp_r \sim \hbar$, leads to an expression for the radial kinetic energy, $\hbar^2/2M_{red}r^2$, which has the same form as the centrifugal barrier, with an "l" value ≈ 0.6. Thus $l = 0$ and $l = 1$ states tend to be close in energy [‡12]. The masses of the quarks and of at least the charged leptons in-

[‡10] Visnjić-Triantafillou [1] has recently discussed this issue in a somewhat different way.

[‡11] In terms of the general charge pattern of footnote 2, the condition that the charged lepton of a given generation be more massive than its neutrino is $q > -1$. The condition that the up-type quark be more massive than the down-type quark for higher generations (ignoring the u–d near equality in mass) is $q < -1/3$. Thus the reversal in the mass relation, $m_c > m_s$, etc. for quarks, and $m_{\nu_\mu} < m_\mu$, etc. for leptons, might be due to electromagnetism;

[‡12] A more radical and speculative way of elevating orbital excitations is to imagine that, phenomenologically, a continuum degree of freedom is present. For example, the QSD gauge flux which permanently confines the quark and lepton constituents may be equivalent to a "rigid" string or "rod". We can then identify the longitudinal excitations of the rod, which have angular momentum zero, with the higher generations, and compare the energy, E_n, associated with longitudinal excitations, with E_l, the energy associated with rotational excitations of the rod which have non-zero angular momentum l. We estimate $E_n = (2n + 1)\pi h v/L$, and $E_l = 5l(l + 1)\hbar^2/2ML^2$, where v is the velocity of longitudinal waves in the rod, and L and m are length and the mass of the rod, respectively. The condition for $E_n \ll E_l$ is $\pi M L v \ll 25\hbar$.

Volume 99B, number 4 PHYSICS LETTERS 26 February 1981

crease rapidly with generation. In the semiclassical approximation for the energy levels E_n of radial states of a non-relativistic particle in a potential $\sim r^{\alpha}$, the most rapid increase with n, $E_n \propto n^2$, occurs for $\alpha \to \infty$, which corresponds to an infinite square well, an opposite limit from the limit $\alpha \to -2$ which allows many radial excitations below the first orbital excitation.

In summary, we have proposed a composite model of quarks, leptons, weak vector bosons, and Higgs mesons which has certain elements of simplicity: most of these particles are two-body composites, most of the two-body composites correspond to desired particles, quark–lepton parallelism holds, and the strong and weak interactions are on a parallel footing – both are residual effects of flavor-independent local gauge interactions. The model accounts, in a schematic way, for the generations of quarks and leptons, and predicts generations of narrow states up to the W mass. Under assumptions stated in the text, the model has Cabibbo structure, GIM mechanism, and spontaneously-broken left–right symmetry for the weak interactions.

Although tests of composite models such as ours generally require energies large enough (at least in the TeV range) to probe the constituents structure of quarks and leptons [‡2], our model does require that there be no narrow-width generations above the W mass.

We are greatly indebted to Jogesh Pati for detailed discussions which resulted in important changes in our point of view. We are grateful to Tony Kennedy and Rabi Mohapatra for a number of helpful remarks. We thank Haim Harari, Shmuel Nussinov, Jon Rosner and Chin-Hung Woo for illuminating conversations. This work was supported in part by the National Science Foundation.

References

[1] O.W. Greenberg and C.A. Nelson, Phys. Rev. D10 (1974) 2567;
O.W. Greenberg, Phys. Rev. Lett. 35 (1975) 1120; and references cited there;
J.C. Pati, A. Salam and J. Strathdee, Phys. Lett. 58B (1975) 265;
E. Nowak, J. Sucher and C.H. Woo, Phys. Rev. D16 (1977) 2874;
H. Harari, Phys. Lett. 86B (1979) 83;
M.A. Shupe, Phys. Lett. 86B (1979) 87;
G. 't Hooft, Lecture III, Cargèse Summer Institute (1979);
H. Terazawa, Phys. Rev. D22 (1980) 184;
W. Królikowski, Phys. Lett. 85B (1979) 335; Acta Phys. Pol. B10 (1979) 739; B11 (1980) 431;
R. Casalbuoni and R. Gatto, Phys. Lett. 93B (1980) 47;
J. Ellis, M.K. Gaillard, L. Maiani and B. Zumino. LAPP preprint TH-15/CERN preprint TH 2481 (1980), to be published in Proc. Europhysics Study Conf. on Unification of fundamental interactions (Erice, 1980);
J. Ellis, M.K. Gaillard, and B. Zumino, Phys. Lett. 94B (1980) 343;
V. Visnjić-Triantafillou, Phys. Lett. 95B (1980) 47; and Fermilab preprint FERMILAB-Pub-80/34-THY (1980);
G.L. Shaw and R. Slansky, Phys. Rev. D22 (1980) 1760;
J.C. Pati, Bhubaneswar and Trieste preprint (1980);
S. Dimopoulos, S. Raby and L. Susskind, Stanford preprint ITP-662 (1980).

[2] S. Weinberg, Phys. Rev. D13 (1976) 974; 19 (1979) 1277;
L. Susskind, Phys. Rev. D20 (1979) 2619;
S. Dimopoulos and L. Susskind, Nucl. Phys. B155 (1979) 237;
S. Dimopoulos, S. Raby and L. Susskind, Stanford preprint ITP-662 (1980);
S. Raby, S. Dimopoulos and L. Susskind, Stanford preprint ITP-635 (1979).

[3] H. Georgi, H. Quinn and S. Weinberg, Phys. Rev. Lett. 33 (1974) 451.

[4] J.C. Pati and A. Salam, Phys. Rev. Lett. 31 (1973) 661; Phys. Rev. D10 (1974) 275;
G. Senjanović and R.N. Mohapatra, Phys. Rev. D12 (1975) 1502;
G. Senjanović, Nucl. Phys. B153 (1979) 334;
and references cited in these articles.

[5] P.Q. Hung and J.J. Sakurai, Nucl. Phys. B143 (1978) 81.

[6] H.J. Lipkin, Phys. Lett. 89B (1980) 358;
G.L. Shaw, D. Silverman and R. Slansky, Los Alamos preprint LA-UR-80-588 (1980);
S. Brodsky and S.D. Drell, SLAC preprint SLAC-PUB-2534 (1980).

[7] C. Quigg and J.L. Rosner, Phys. Rep. 56 (1979) 167.

Reprinted with permission from *Physics Letters,* **124B** (1, 2), pp. 67–73, W. Buchmuller, R.D. Peccei, and T. Yanagida, "Quarks and Leptons as Quasi Nambu-Goldstone Fermions." © 1983, North Holland Publishing Company.

Volume 124B, number 1,2 PHYSICS LETTERS 21 April 1983

QUARKS AND LEPTONS AS QUASI NAMBU–GOLDSTONE FERMIONS

W. BUCHMÜLLER, R.D. PECCEI and T. YANAGIDA [1]

Max-Planck-Institut für Physik und Astrophysik – Werner Heisenberg Institut für Physik – Munich, Fed. Rep. Germany

Received 15 December 1982

Dedicated to the memory of J.J. Sakurai

We discuss a new idea for constructing composite quarks and leptons which have (approximately) vanishing mass. They are associated with fermionic partners of Goldstone bosons arising from the spontaneous breakdown of an internal symmetry G_f in a supersymmetric preon theory. For $G_f = SU(5)$ being broken to $SU(3) \times U(1)_{em}$ there arise as quasi Goldstone fermions, naturally and unequivocally, precisely the quarks and leptons of one family. The dynamics of these quasi Goldstone fermions is explored by constructing a general supersymmetric nonlinear effective lagrangian. By means of a reduced model, we show that the first nontrivial interactions of the quasi Goldstone fermions can give rise, in an effective way, to the weak interactions. Issues connected with the incorporation of families in the scheme and the generation of masses, as well as the possible structure of the underlying preon theory are briefly discussed.

At present we have no evidence that quarks and leptons are anything but elementary excitations. Yet there has been a growing interest recently in considering models of composite quarks and leptons [1]. This is motivated principally by the feeling that, unless quarks and leptons are composite, there is no hope in ever understanding the complicated patterns of their masses. Composite models of quarks and leptons (preon theories) face an immediate challenge: they must explain why the resulting bound states have sizes, $\langle r \rangle \sim 1/\Lambda$, which are so much smaller than their Compton wavelengths, $1/m$. The strong inequality $\Lambda \gg m$ is only understandable dynamically if there exists an, approximate, symmetry which forces certain bound states to have zero mass. It was originally suggested by 't Hooft [2] that such a symmetry might be chiral symmetry. However, the construction of realistic models of quarks and leptons based on this idea is made difficult by the necessity of matching chiral anomalies at the preon and composite level. Although models exist in which these anomalies do match [3], these models tend to be rather complicated, often involving more preons than quarks and leptons.

The purpose of this note is to discuss an alternative possibility for obtaining light composite quarks and leptons. The dynamical mechanism we suggest to produce (approximately) massless bound states makes use of supersymmetry and of the spontaneous breakdown of an internal symmetry, rather than of chiral symmetry. Our mechanism is the following: imagine a preon theory which is supersymmetric and which has an overall global internal symmetry group G_f. If this global symmetry is spontaneously broken down to a subgroup H_f, then there will appear in the theory Goldstone bosons – one for each of the generators in G_f/H_f. However, because of the supersymmetry, these Goldstone bosons are accompanied by fermionic partners – which in an earlier publication with Love [4] we called quasi Goldstone fermions – as well as an additional set of bosonic partners. Our suggestion is then simply that quarks and leptons are quasi Goldstone fermionic bound states of a confining supersymmetric theory [‡1].

[1] On leave of absence from the Department of Physics, College of General Education, Tohoku University, Sendai, Japan.

[‡1] Our asking that quark and leptons are bound states of a confining theory, rather than elementary excitations is tied to our desire of having the possibility of eventually calculating their masses. The confinement requirement for the preon theory follows only because it is hard to imagine confined quarks made out of unconfined preons!

0 031-9163/83/0000–0000/$ 03.00 © 1983 North-Holland

Volume 124B, number 1,2 PHYSICS LETTERS 21 April 1983

This idea has two apparent drawbacks, which at first sight may appear to render it phenomenologically useless:

(1) The quasi Goldstone fermions are accompanied by partner scalar excitations. One knows experimentally that the scalar partners of quarks and leptons, if they exist, must be considerably heavier than the fermions.

(2) Quasi Goldstone fermions transform as real or as pairs of complex conjugate representations under the unbroken group H_f. We know, however, that with respect to the weak interactions the quarks and leptons are complex and not paired.

The first point above is a common problem of any "low energy" supersymmetric theory, which may be in principle circumventable by an appropriate breaking of the supersymmetry. The second point, however, appears much more serious, and in fact impeded progress for some time.

There is an advantage of our scheme. Namely, since G_f and H_f are at our disposal, it may be possible to select them appropriately to obtain quarks and leptons with the right quantum numbers. For instance [4], with $G_f = E_6$ and $H_f = SO(10)$ one finds in G_f/H_f a $16 + \overline{16} + 1$ set of massless states. Thus, the quarks and leptons of one family can be fit in the 16 but, illustrating point (2) above, there is an additional $\overline{16}$ of mirror fermions [‡2].

The impasse surrounding the mirroring of representations can, however, be overcome. Let us focus for the moment on one family of 15 quarks and leptons (15 left handed fields). Although this set of states is not real with respect to the standard model group $SU(3) \times SU(2) \times U(1)$, *it is real with respect to the unbroken subgroup of* $SU(3) \times U(1)_{em}$. That is, all the quark and lepton representations come in complex conjugate pairs, save for the neutrino which is a singlet under $SU(3) \times U(1)_{em}$. Thus if H_f is $SU(3) \times U(1)_{em}$, it is perfectly reasonable to suppose that the 15 quarks and leptons of one generation are the quasi Goldstone fermions of the breakdown of $G_f \to H_f$. Since $9 + 15 = 24$, clearly the unique solution for G_f is SU(5)!

It may be useful to restate this result. If one has a supersymmetric preon theory which has an SU(5) global symmetry which is spontaneously broken down to $SU(3) \times U(1)_{em}$, necessarily and unequivocally, one obtains a set of 15 massless fermionic excitations with precisely the quantum numbers of the quarks and leptons of one family. We note that this SU(5) has an electromagnetic charge definition which is different from the usual Georgi–Glashow SU(5) [7]. Namely for our case

$$Q = \mathrm{diag}\left(\tfrac{1}{15}, \tfrac{1}{15}, \tfrac{1}{15}, \tfrac{2}{5}, -\tfrac{3}{5}\right). \qquad (1)$$

Using (1) it is easy to check that the Goldstone excitations which are in $SU(5)/SU(3) \times U(1)_{em}$ have precisely the charges and color quantum numbers of the quarks and leptons.

If quarks and leptons are the quasi Goldstone fermions of $SU(5)/SU(3) \times U(1)_{em}$ it is possible to discover some of their dynamical interactions, even though we do not really know specific details of the preon theory. This has an analog in hadronic physics. Even before one knew about QCD, low energy pion physics was understood because the pions were the Goldstone bosons of the spontaneous breakdown of chiral $SU(2) \times SU(2)$. The Goldstone nature of the pions allowed for the construction of effective lagrangians [8] where the chirality is realized nonlinearly. These lagrangians embody all the low energy properties of the theory. For the case at hand, we need a supersymmetric generalization of this construction applied to the specific groups G_f and H_f in question.

Before we outline the details of this construction, it will be helpful to indicate the nature of the results obtained. We find that the dominant interaction at low energy among the quasi Goldstone fermions is a four-Fermi current–current interaction, which in general contains a series of parameters characterizing the breakdown. For the case of a particularly simple toy model – to be described below – these parameters can be chosen *to reproduce precisely the current–current form of the weak interactions.* We expect that this behaviour should also persist in the more realistic $SU(5) \to SU(3) \times U(1)_{em}$ breakdown. This result suggests – although it cannot prove – that if quarks and leptons are indeed quasi Goldstone fermions then the weak interactions may be residual. Before discussing this point further, as well as possible ideas for mass generation and the inclusion of families in

[‡2] For grassmanian manifolds of dimension $2M$ there is the intriguing possibility of getting rid of the doubling of fermions by supersymmetrizing only M complex scalar fields. See refs. [5] and [6].

Volume 124B, number 1,2 PHYSICS LETTERS 21 April 1983

our scheme, we want to proceed to the construction of the effective lagrangian which yields the above quoted result.

What is needed is a generalization of the Coleman, Wess and Zumino [9] construction of nonlinear lagrangians applied to the case of supersymmetric theories. Although some such constructions exist already in the literature [10,11], they do not really quite fit our needs. We outline therefore a general procedure which we have found expiditious for constructing these nonlinear supersymmetric lagrangians. For a given group G_f with generators T_α one can always construct nonlinear realization of the Goldstone fields ξ_i in the coset G_f/H_f by explicitly constructing the Killing vectors $A_{\alpha i}(\xi)$ [‡3]. The commutators of the generators T_α with the Goldstone field ξ_i are then given by

$$i^{-1}[T_\alpha, \xi_i] = A_{\alpha i}(\xi), \tag{2}$$

and one can check that for indices α corresponding to generators in H_f the transformation is linear, otherwise it is nonlinear. The construction in (2) automatically satisfies all the Jacobi identities. It is this property that makes it immediately useful for a supersymmetric generalization. Let ϕ_i be a chiral superfield which contains the ξ_i as part of its scalar components. Then the equation

$$i^{-1}[T_\alpha, \phi_i] = A_{\alpha i}(\phi) \tag{3}$$

obviously provides an adequate nonlinear supersymmetric realization, since it also satisfies all the required Jacobi identities. This construction is the analog of what was done by Weinberg [8] originally for chiral symmetry.

Armed with (3), it is now straightforward to construct a nonlinear lagrangian that is invariant under these transformations. First of all we note that in (3) there will appear in general arbitrary scales V_i – one such scale for each of the different representations of H_f under which the Goldstone superfields ϕ_i transform [12] [‡4]. Focusing only on the nonlinear part (generators T_i) we may write therefore (3) in the form

$$i^{-1}[T_i, \phi_i/V_j] = C_{ij} + C_{ijk}\,\phi_k/V_k + C_{ijkl}(\phi_k/V_k)\phi_l/V_l, \tag{4}$$

where the coefficients C_{ij}, C_{ijk}, etc. are numbers determined by group theory. If we are interested in an effective nonlinear lagrangian invariant under (4) up to and including interactions of $O(1/V^N)$, we need to compute (4) only up to terms of similar order. For our purposes it will suffice to stop at terms of $O(1/V^2)$.

Our lagrangian must include, of course, the generalized kinetic energy for all the Goldstone superfields, $(\phi_i^\dagger \phi_i + ...)_{D\text{-term}}$, invariant under (4). It may, furthermore, include any higher order interactions among the superfield ϕ_i which are: (a) supersymmetric and (b) invariant themselves under (4) [‡5]. It is easy to check that *only D*-terms are allowed. *F*-term interactions necessarily require that some of the scalar components of the ϕ_i have non-zero vacuum expectation values – thereby contradicting the Goldstone nature of these fields. We note furthermore that, given a nonlinear realization, one has the freedom to reparametrize the fields again nonlinearly without affecting the physics. Thus, using this freedom, it is easy to convince oneself that, up to terms of $O(1/V^3)$, the most general nonlinear supersymmetric lagrangian has the form

$$\mathcal{L} = \phi_i^\dagger \phi_i + A_{ijkl}(\phi_i^\dagger/V_i)(\phi_j^\dagger/V_i)(\phi_k/V_k)\phi_l/V_l + \text{h.c.}|_{D\text{-term}}, \tag{5}$$

where the coefficients A_{ijkl} have dimensions of mass squared. In general these coefficients will contain pieces proportional to the various scales V_i, as well as pieces corresponding to new invariant tensor structures.

‡3 The paper of Boulware and Brown [12] presents a lucid discussion of the general construction of the Killing vectors $A_{\alpha i}(\xi)$ and of nonlinear realizations.

‡4 For the case of chiral symmetry the pions transform as vectors under the unbroken isospin group and hence there is only one dimensionful constant f_π.

‡5 This freedom arises because the lagrangian corresponds to the Kahler potential of a complex manifold whose metric is not uniquely determined by the Killing vectors of G/H.

Volume 124B, number 1,2 PHYSICS LETTERS 21 April 1983

The first nonlinear correction to the Goldstone superfield kinetic energies contains precisely the four-Fermi current–current interactions discussed earlier. Using a Fierz identity and writing only the piece involving fermions (written as left-handed Dirac fields) one has that

$$\phi_i^\dagger \phi_j^\dagger \phi_k \phi_l|_{D\text{-term}} = -\tfrac{1}{2}(\bar{\psi}_{iL}\gamma^\mu \psi_{kL})(\bar{\psi}_{jL}\gamma_\mu \psi_{lL}) + \text{terms involving boson fields.} \tag{6}$$

Hence (5) does indeed contain a vector (or axial vector) interaction between currents formed out of the quasi Goldstone fermions. We should note here the difference between quasi Goldstone fermions and goldstinos. If the supersymmetry were spontaneously broken one would also obtain four-fermion interactions among the goldstinos. However, the coupling would involve not $\gamma^\mu \gamma_\mu$ but $\gamma^\mu \partial^\nu \gamma_\mu \partial_\nu$, which vanishes as the momentum transfer vanishes. Goldstinos can never reproduce the weak interactions [13], but quasi Goldstones may [‡6]!

The construction of the nonlinear lagrangian (3) is perfectly straightforward but for the case of the breakdown $SU(5) \to SU(3) \times U(1)_{em}$ it is rather laborious. In this note, therefore, we will discuss only a toy example involving the breakdown of $U(2) \to U(1)_{em}$. This "model for leptons" will illustrate most of the relevant physics. We reserve for a forthcoming publication [16] the detailed discussion of the $SU(5) \to SU(3) \times U(1)_{em}$ case.

Let us denote the broken generators of U(2) by T_+, T_- and T_0 and let the corresponding Goldstone superfields be ϕ_+, ϕ_- and ϕ_0, where the subscripts give the relevant EM charges. The nonlinear realization corresponding to the $U(2)/U(1)_{em}$ breakdown will give rise to the commutators

$$i^{-1}[T_i, \phi_j] = A_{ij}(\phi), \quad i, j = +, -, 0. \tag{7}$$

A straightforward calculation, using the methods of ref. [12] to construct the Killing vectors, gives, up to terms of $O(1/V^3)$, the following formula for the matrix A_{ij}:

$$A_{ij} = -\begin{array}{c|ccc} & + & - & 0 \\ \hline + & \frac{5}{24}\phi_+\phi_+/V_+ & \begin{array}{l} -(1/2i)(V_0/V_+)\phi_+ \\ +\frac{1}{24}\phi_0\phi_+/V_+ \end{array} & \begin{array}{l} 2V_- + (1/2i)(V_-/V_0)\phi_0 \\ -\frac{1}{24}(V_-/V_0^2)\phi_0^2 - \frac{5}{24}\phi_+\phi_-/V_+ \end{array} \\ - & (1/2i)\phi_+ + \frac{1}{24}\phi_+\phi_0/V_0 & \begin{array}{l} 2V_0 \\ -\frac{2}{24}(V_0/V_+V_-)\phi_+\phi_- \end{array} & -(1/2i)\phi_- + \frac{1}{24}\phi_0\phi_-/V_0 \\ 0 & \begin{array}{l} 2V_+ - (1/2i)(V_1/V_0)\phi_0 \\ -\frac{1}{24}(V_+/V_0^2)\phi_0^2 - \frac{5}{24}\phi_+\phi_-/V_- \end{array} & \begin{array}{l} (1/2i)(V_0/V_-)\phi_- \\ +\frac{1}{24}\phi_0\phi_-/V_- \end{array} & \frac{5}{24}\phi_-\phi_-/V_- \end{array} . \tag{8}$$

Having eq. (8) in hand, it is now easy but tedious to construct an effective lagrangian which is invariant under these transformations, up to terms of $O(1/V^3)$. The result of this computation is the following nonlinear lagrangian, in which some simplifications were made by using the freedom of arbitrarily reparametrizing the superfields ϕ_i:

$$\begin{aligned} \mathcal{L}_{eff} = {} & \phi_+^\dagger\phi_+ + \phi_-^\dagger\phi_- + \phi_0^\dagger\phi_0 + \{\tfrac{1}{4}(A + B - \tfrac{1}{2}V_+^2 + \tfrac{3}{16}V_0^2)\phi_+^\dagger\phi_+^\dagger\phi_+\phi_+/V_+^4 + \tfrac{1}{4}(A - B - \tfrac{1}{2}V_-^2 + \tfrac{3}{16}V_0^2)\phi_-^\dagger\phi_-^\dagger\phi_-\phi_-/V_-^4 \\ & + (D - C^2/V_0^2)\phi_0^\dagger\phi_0^\dagger\phi_0\phi_0/V_0^4 + [2E + \tfrac{1}{2}C - V_+^{-2}(B + \tfrac{1}{4}V_0^2)^2]\,\phi_+^\dagger\phi_0^\dagger\phi_+\phi_0/V_+^2V_0^2 \\ & + (A - B^2/V_0^2)\phi_+^\dagger\phi_-^\dagger\phi_+\phi_-/V_+^2V_-^2 + [2E - \tfrac{1}{2}C - V_-^{-1}(B - \tfrac{1}{4}V_0^2)^2]\,\phi_-^\dagger\phi_0^\dagger\phi_-\phi_0/V_-^2V_0^2 \\ & + (E - BC/V_0^2)(\phi_+^\dagger\phi_-^\dagger\phi_0\phi_0 + \phi_0^\dagger\phi_0^\dagger\phi_+\phi_-)/V_+V_-V_0^2\} + O(1/V^3)|_{D\text{-term}} . \end{aligned} \tag{9}$$

‡6 In the recent suggestion of Bardeen and Višnjić [14] that quarks and leptons are the goldstinos of spontaneously broken supersymmetric theories (see also ref. [15]) the resulting nonlinear derivative interactions are interpreted as characteristic signals of the onset of compositeness.

Here the constants A, B, C, D, E, with dimension of $(\text{mass})^2$ are arbitrary and characterize additional invariants allowed under the transformation (7). The terms involving only the scales V_i give the invariant generalization of the kinetic energy to this order.

Because of the rather large amount of freedom in eq. (9), it is not surprising that such a lagrangian can reproduce the effective current–current form of the standard model. In fact, for $\sin^2\theta_W = \frac{1}{4}$ and $\rho = 1$ for simplicity, the standard model lagrangian in units of $\sqrt{2}\,G_F$ necessitates the following coefficients for the terms appearing in eq. (9): $-\frac{1}{4}, -\frac{1}{4}, -1, +1, -3, -\frac{1}{2}, 0$. It is easy to convince oneself that such numbers can be obtained via suitable choices of $A, ..., E$ and V_+, V_- and V_0.

The lessons to be learned from this model depend in part on the amount of speculation one allows oneself. It is clear that, although we have shown that the residual dynamics are *compatible* with the standard four-fermion weak interactions, it could well be that these interactions have nothing to do with weak processes at all. Indeed if the $V_i \gg 250$ GeV then these interactions could just be additional short range contributions which slightly modify the normal weak interactions (tell tale signs of compositeness?). On a much more speculative level, however, is the possibility that the weak interactions *are* these residual interactions. Obviously, the particular structure of the coefficients $A, ..., E$ and V_i's which give one this option must reflect particular dynamical properties of the underlying theory. If one has residual weak interactions then the Z^0 and $W^{\pm}$ of the standard model, although probably present dynamically, would have different properties (masses, widths, number of states, etc.). In this sense this idea is very much along the lines pursued by Sakurai [17]. Note, in particular, that if the interactions of eq. (9) are the weak interactions, then the scale of the weak interactions and that of compositeness are essentially the same. This proposition suggests therefore a nearby compositeness with a concomitant rich and interesting physics.

The idea that quarks and leptons are quasi Goldstone fermions clearly opens up interesting vistas. However, there are many open and interesting problems which need to be urgently answered. Foremost among these are:

(1) How does one incorporate families?

(2) How does one generate masses?

(3) What is the structure of the underlying preon theory?

We obviously do not have very precise ideas on how to answer these universal and deep questions. Nevertheless, we want to close this paper by outlining some of our tentative thoughts on these issues.

Families have probably a deeper reason for their existence. However, if one just wanted to incorporate them in the quasi Goldstone way as group theoretical repetitions, then there are not that many possibilities because the group theory is quite restrictive. The two more "natural" ways we have found are:

(1) $G_f = (SU(5))^{n_f}$ broken down to $(SU(3) \times U(1))^{n_f}$, with the color and $U(1)_{em}$ group being eventually identified with the diagonal sums; or:

(2) $G_f = E_7$ broken down to $H_f = SU(6) \times SU(3)$.

In this last case the coset space G_f/H_f has 90 excitations which comprise 3 families of states plus their mirrors. The mirroring here may ultimately prove to be unacceptable, but the fact that E_7 is the global symmetry connected with $N = 8$ supergravity [18] makes this alternative worth exploring.

The issue of mass generation is most likely connected to the existence of families. Nevertheless, perhaps it is worthwhile describing here some of our intuitive ideas, even within the context of one family. The picture we have is the following: Goldstone excitations can acquire small masses when the global symmetry, whose spontaneous breakdown engendered them, is slightly broken. The classical application of this idea is in pion physics: The spontaneous breakdown of $SU(2) \times SU(2)$ gives three massless pions. However, turning on α_{em} breaks explicitly the $SU(2) \times SU(2)$ and the $\pi^{\pm}$ get a mass shift relative to the π^0. This shift $\delta\, m^2_{\pi^{\pm}}$ in QCD is obviously of $O(\alpha\Lambda^2_{QCD})$. Indeed, even before the advent of QCD it was calculated [19] in terms of appropriately saturated vector and axial vector spectral functions and related to α and the ρ and A_1 masses.

We expect that the SU(5) global symmetry of the preon theory is explicitly broken when we turn on the SU(3) and $U(1)_{em}$ coupling constants. Hence, intuitively, the electron should get a mass because it has charge; the quarks get masses because they carry color and charge, and the neutrino remains massless. This expectation,

Volume 124B, number 1,2 PHYSICS LETTERS 21 April 1983

however, is not borne out if the supersymmetry remains exact! This is because of the, so called, nonrenormalization theorem [20] of supersymmetric theories, which does not allow particles which have zero mass at tree level to acquire radiative masses ‡7. For mass to be generated for the Goldstone bosons and the quasi Goldstone fermions it is necessary that the supersymmetry be broken first.

It is an open question how the supersymmetry is actually to be broken. One option which might be useful to consider, at least for the purpose of mass generation, is the following. Imagine that the gauging of the SU(3) $\times$ $U(1)_{em}$ is done in a nonsupersymmetric way, by omitting altogether the gauginos ‡8. In this case, one expects that the scalar excitations pick up a $(\text{mass})^2$ of $O(\alpha\Lambda_{MC}^2)$ where Λ_{MC} is the typical scale associated with the compositeness. This is very much like in the pion case. The quasi Goldstone fermions cannot, however, get a mass directly in this way, since massless fermions are chiral. Their mass can only arise through graphs in which the quasi Goldstone fermions couple directly with the heavy states in the theory. In zeroth order, these graphs themselves cannot generate mass, but eventually, through the dressing with the nonsupersymmetric gauge fields some mass can be expected. It is obviously hard at this stage to give an estimate of the dynamical mass so acquired by the quasi Goldstone fermions. However, it appears certain that it will be less than that of the corresponding bosons. But will it be light enough to satisfy the necessary phenomenological requirements?

The question of the structure of the preon theory is, obviously, the hardest to answer, and here we are likely to be wrong. A possibility which we have been exploring is an $SU(5)_{MC}$ of metacolor with 5_L and 5_R flavors of preons. This theory requires a mass term for the flavor preons so as to have only a vectorial global SU(5). The natural $SU(5)_{MC}$ singlet condensates $\langle 5\bar{5}\rangle$ and $\langle 55555\rangle$, $\langle\bar{5}\bar{5}\bar{5}\bar{5}\bar{5}\rangle$ can break the SU(5) of flavor (via a 24 plus a 175 and $\overline{175}$) to $SU(3)\times U(1)_{em}$. In this theory it is possible to identify an SU(2) which could correspond to the weak interactions among the lower two components of the 5_L preons ‡9. The open question, however, is whether one should gauge these interactions then at the preonic level or make use of the bound state dynamics to generate the weak interactions via the particular pattern of symmetry breakdown. It is clear that this and many other open questions will need to await future developments for their answer.

We have profited greatly from conversations with D. Amati, P. Breitenlohner, A. Buras, C.K. Lee, L. Girardello and D. Maison. We are also grateful to many colleagues, both at the Max-Planck-Institut as well as elsewhere, for their thoughtful criticisms. We are sorry that because of S. Love's return to the USA he could not participate in these new developments.

‡7 One should perhaps worry about applying these theorems to bound states, but it appears to us that they probably should also go through in this case.

‡8 Such an option is not sensible if the Goldstone excitations are elementary, since it breaks the supersymmetry in a hard way. It might be sensible for composite Goldstone particles, but it must be modified at the preon level. At that level it is more reasonable to add a gaugino mass.

‡9 Note, however, that the mass term $M\,\bar{5}_L 5_R$ explicitly breaks this SU(2).

References

[1] For a review see for example: R.D. Peccei, Lectures Arctic School of Physics (Äkäslompolo, Finland) MPI-PAE/PTh 69/82.
[2] G. 't Hooft, in: Recent developments in gauge theories, Cargèse Lectures (1979) (Plenum, New York, 1980) p. 135.
[3] I. Bars, Nucl. Phys. B208 (1982) 77.
[4] W. Buchmüller, S.T. Love, R.D. Peccei and T. Yanagida, Phys. Lett. 115B (1982) 233.
[5] C.L. Ong, SLAC preprint SLAC-PUB-2936 (1982).
[6] J. Bagger and E. Witten, Phys. Lett. 118B (1982) 103.
[7] H. Georgi and S.L. Glashow, Phys. Rev. Lett. 32 (1974) 438.
[8] S. Weinberg, Phys. Rev. 166 (1968) 1568.
[9] S. Coleman, J. Wess and B. Zumino, Phys. Rev. 177 (1969) 2139;
C.G. Callan, S. Coleman, J. Wess and B. Zumino, Phys. Rev. 177 (1969) 2247.

Volume 124B, number 1,2 PHYSICS LETTERS 21 April 1983

[10] B. Zumino, Phys. Lett. 87B (1979) 203.
[11] L. Alvarez-Gaumé and D.Z. Freedman, Commun. Math. Phys. 80 (1981) 443.
[12] D.G. Boulware and L.S. Brown, Ann. Phys. (NY) 138 (1982) 392.
[13] B. de Wit and D.Z. Freedman, Phys. Rev. Lett. 35 (1975) 827;
W.A. Bardeen, unpublished.
[14] W.A. Bardeen and V. Višnjić, Nucl. Phys. B194 (1982) 422.
[15] H. Terazawa, Prog. Theor. Phys. 64 (1980) 1763.
[16] W. Buchmüller, R.D. Peccei and T. Yanagida, in preparation.
[17] J.J. Sakurai, Schladming Lectures (1982).
[18] E. Cremmer and B. Julia, Phys. Lett. 80B (1978) 48; Nucl. Phys. B159 (1979) 141.
[19] T. Das, G.S. Guralnik, V.S. Mathur, F.E. Low and J. Young, Phys. Rev. Lett. 18 (1967) 759
[20] J. Iliopoulos and B. Zumino, Nucl. Phys. B76 (1974) 310;
S. Ferrara, J. Iliopoulos and B. Zumino, Nucl. Phys. B77 (1974) 413.

VIOLATION OF CP INVARIANCE, C ASYMMETRY, AND BARYON ASYMMETRY OF THE UNIVERSE

A. D. Sakharov

Submitted 23 September 1966
ZhETF Pis'ma 5, No. 1, 32-35, 1 January 1967

The theory of the expanding Universe, which presupposes a superdense initial state of matter, apparently excludes the possibility of macroscopic separation of matter from antimatter; it must therefore be assumed that there are no antimatter bodies in nature, i.e., the Universe is asymmetrical with respect to the number of particles and antiparticles (C asymmetry). In particular, the absence of antibaryons and the proposed absence of baryonic neutrinos implies a non-zero baryon charge (baryonic asymmetry). We wish to point out a possible explanation of C asymmetry in the hot model of the expanding Universe (see [1]) by making use of effects of CP invariance violation (see [2]). To explain baryon asymmetry, we propose in addition an approximate character for the baryon conservation law.

We assume that the baryon and muon conservation laws are not absolute and should be unified into a "combined" baryon-muon charge $n_c = 3n_B - n_\mu$. We put:

$n_\mu = -1$, $n_K = +1$ for antimuons μ_+ and $\nu_\mu = \mu_0$,

$n_\mu = +1$, $n_K = -1$ for muons μ_- and $\nu_\mu = \mu_0$,

$n_B = +1$, $n_K = +3$ for baryons P and N,

$n_B = -1$, $n_K = -3$ for antibaryons P and N

This form of notation is connected with the quark concept; we ascribe to the p, n, and λ quarks $n_c = +1$, and to antiquarks $n_c = -1$. The theory proposes that under laboratory conditions processes involving violation of n_B and n_μ play a negligible role, but they were very important during the earlier stage of the expansion of the Universe.

We assume that the Universe is neutral with respect to the conserved charges (lepton, electric, and combined), but C-asymmetrical during the given instant of its development (the positive lepton charge is concentrated in the electrons and the negative lepton charge in the excess of antineutrinos over the neutrinos; the positive electric charge is concentrated in cesses that include the a-boson only virtually.

We estimate the constant of this interaction at $g_a = 137^{-3/2}$, from the following considerations: The vector interaction of the a-boson with the μ-neutrino leads to the presence of a certain rest mass in the latter. The upper bound of the mass μ_0 is estimated in [7] on the basis of cosmological considerations. If we assume a flat cosmological model of the Universe and assume that the greater part of its density $\rho \sim 1.2 \times 10^{-29}$ g/cm^3 should be ascribed to μ_0, then the rest mass of μ_0 is close to 30 eV. The given value of g_a follows then from the hypothetical formula

$$\frac{m_{\mu_0}}{m_e} = \frac{g_a^2}{e^2} \sim (137)^{-2}.$$

We note that the presence in the Universe of a large number of μ_0 with finite rest mass should lead to a number of very important cosmological consequences.

2. The baryon charge is violated if the interaction described in Item 1 is supplemented with a three-boson interaction leading to virtual processes of the type $a_{\alpha_1} + a_{\alpha_2} + a_{\alpha_3} \to 0$. At the advice of B. L. Ioffe, I. Yu. Kobzarev, and L. B. Okun', the Lagrangian of this interaction is assumed to be dependent on the derivatives of the a-field, for example,

$$L_2 = g_2(\sum_\alpha f_k^i f_j^k f_i^j + \text{h.c.}),\quad f_{ik} = R_0 t a_i.$$

Inasmuch as L_2 vanishes when two tensors coincide, in this concrete form of the theory we should assume the presence of several types of a-fields. Assuming $g_2 = 1/M_0^2$ and $M_0 = 2 \times 10^{-5}$ g, we have strong interaction at $n \sim 10^{98}$ cm^{-3} and very weak interaction under laboratory conditions. The figure shows a proton-decay diagram including three vertices of the first type, one vertex of the second, and the vertex of proton decay into quarks, which we assume to contain the factor $1/p_q^2$ (due, for example, to the propagator of the "diquark" boson binding the quarks in the baryon). Cutting off the logarithmic divergence at $p_q = M_0$, we find the decay probability

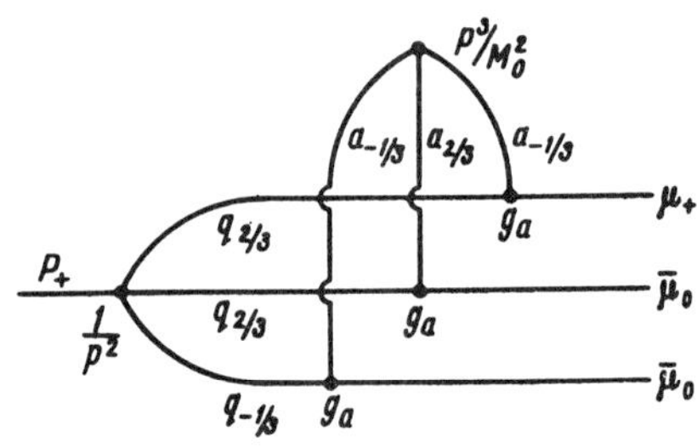

$$\omega \sim \frac{m_p^5 g_a^6 [\ln(M_0/m_a)]^2}{M_0^4}.$$

The lifetime of the proton turns out to be very large (more than 10^{50} years), albeit finite.

The author is grateful to Ya. B. Zel'dovich, B. Ya. Zel'dovich, B. L. Ioffe, I. Yu. Kobzarev, L. B. Okun', and I. E. Tamm for discussions and advice.

[1] Ya. B. Zel'dovich, UFN 89, 647 (1966) (Review), Soviet Phys. Uspekhi 9 (1967), in press.
[2] L. B. Okun', UFN 89, 603 (1966) (Review), Soviet Phys. Uspekhi 9 (1967), in press.
[3] M. A. Markov, JETP 51, 878 (1966), Soviet Phys. JETP 24 (1967), in press.

the protons and the negative in the electrons; the positive combined charge is concentrated in the baryons, and the negative in the excess of μ-neutrinos over μ-antineutrinos).

According to our hypothesis, the occurrence of C asymmetry is the consequence of violation of CP invariance in the nonstationary expansion of the hot universe during the superdense stage, as manifest in the difference between the partial probabilities of the charge-conjugate reactions. This effect has not yet been observed experimentally, but its existence is theoretically undisputed (the first concrete example, Σ_+ and Σ_- decay, was pointed out by S. Okubo as early as in 1958) and should, in our opinion, have an important cosmological significance.

We assume that the asymmetry has occurred in an earlier stage of the expansion, in which the particle, energy, and entropy densities, the Hubble constant, and the temperatures were of the order of unity in gravitational units (in conventional units the particle and energy densities were $n \sim 10^{98}$ cm^{-3} and $\epsilon \sim 10^{114}$ erg/cm^3).

M. A. Markov (see [5]) proposed that during the early stages there existed particles with maximum mass on the order of one gravitational unit ($M_0 = 2 \times 10^{-5}$ g in ordinary units), and called them maximons. The presence of such particles leads unavoidably to strong violation of thermodynamic equilibrium. We can visualize that neutral spinless maximons (or photons) are produced at $t < 0$ from contracting matter having an excess of antiquarks, that they pass "one through the other" at the instant $t = 0$ when the density is infinite, and decay with an excess of quarks when $t > 0$, realizing total CPT symmetry of the Universe. All the phenomena at $t < 0$ are assumed in this hypothesis to be CPT reflections of the phenomena at $t > 0$. We note that in the cold model CPT reflection is impossible and only T and TP reflections are kinematically possible. TP reflection was considered by Milne, and T reflection by the author; according to modern notions, such a reflection is dynamically impossible because of violation of TP and T invariance.

We regard maximons as particles whose energy per particle ϵ/n depends implicitly on the average particle density n. If we assume that $\epsilon/n \sim n^{-1/3}$, then ϵ/n is proportional to the interaction energy of two "neighboring" maximons $(\epsilon/n)^2 n^{1/3}$ (cf. the arguments in [6]). Then $\epsilon \sim n^{2/3}$ and $R_0^0 \sim (\epsilon + 3p) = 0$, i.e., the average distance between maximons is $n^{-1/3} \sim t$. Such dynamics are in good agreement with the concept of CPT reflection at the point $t = 0$.

We are unable at present to estimate theoretically the magnitude of the C asymmetry, which apparently (for the neutrino) amounts to about $[(\bar{\nu} - \nu)/(\bar{\nu} + \nu)] \sim 10^{-8} - 10^{-10}$.

The strong violation of the baryon charge during the superdense state and the fact that the baryons are stable in practice do not contradict each other. Let us consider a concrete model. We introduce interactions of two types.

1. An interaction between the quark-muon transformation current and the vector boson field $a_{i\alpha}$, to which we ascribe a fractional electric charge $\alpha = \pm 1/3, \pm 2/3, \pm 4/3$ and a mass $m_a \sim (10 - 10^3) m_p$. This interaction produces reactions $q \to a + \bar{\mu}$, $q + \mu \to a$, etc. The interaction of the first type conserves the fractional part of the electric charge and therefore the actual number of quarks minus the number of antiquarks ($= 3n_B$) is conserved in pro-

[4] A. D. Sakharov, JETP Letters 3, 439 (1966), transl. p. 288.

[5] Ya. B. Zel'dovich and S. S. Gershtein, JETP Letters 4, 174 (1966), transl. p. 120.

VOLUME 41, NUMBER 5 PHYSICAL REVIEW LETTERS 31 JULY 1978

Unified Gauge Theories and the Baryon Number of the Universe

Motohiko Yoshimura
Department of Physics, Tohoku University, Sendai 980, Japan
(Received 27 April 1978)

I suggest that the dominance of matter over antimatter in the present universe is a consequence of baryon-number–nonconserving reactions in the very early fireball. Unified guage theories of weak, electromagnetic, and strong interactions provide a basis for such a conjecture and a computation in specific SU(5) models gives a small ratio of baryon- to photon-number density in rough agreement with observation.

It is known that the present universe is predominantly made of matter, at least in the local region around our galaxy, and there has been no indication observed[1] that antimatter may exist even in the entire universe. I assume here that in our universe matter indeed dominates over antimatter, and I ask within the framework of the standard big-bang cosmology[2] how this evolved from an initially symmetric configuration, namely an equal mixture of baryons and antibaryons. Since the baryon number is not associated with any fundamental principle of physics,[3] such an initial value seems highly desirable. I find in this paper that generation of the required baryon number is provided by grand unified gauge theories[4] of weak, electromagnetic, and strong interactions, which predict simultaneous violation of baryon-number conservation and *CP* invariance. More interestingly, my mechanism can explain why the ratio of the baryon- to the photon-number density in the present universe is so small, roughly of the order[2] of 10^{-8}–10^{-10}.

The essential point of my observation is that in the very early, hot universe the reaction rate of baryon-number–nonconserving processes, if they exist, may be enhanced by extremely high temperature and high density. In gauge models discussed below, the relevant scale of temperature is given by the grand unification mass around 10^{16} GeV where fundamental constituents, leptons and quarks, begin to become indistinguishable. This mass is high enough to make futile virtually all attempts to observe proton decay in the present universe: proton lifetime $\gg 10^{30}$ year.[5] Instead, if my mechanism works, we may say that a fossil of early grand unification has remained in the form of the present composition of the universe.

The laws obeyed by the hot universe at temperatures much above a typical hadron mass (~1 GeV) might, at first sight, appear hopelessly complicated because of many unknown aspects of hadron dynamics. Recent developments of high-energy physics, however, tell that perhaps the opposite is the case. At such high temperatures and densities hadrons largely overlap and an appropriate description of the system is given in terms of pointlike objects—quarks, gluons, leptons, and any other fundamentals. The asymptotic freedom[6] of the strong interaction and weakness of the other interactions further assure[7] that this hot universe is essentially in a thermal equilibrium state made of almost freely moving objects. I shall assume that this simple picture of the universe is correct up to a temperature close to the Planck mass, $G_N^{-1/2} \sim 10^{19}$ GeV, except possibly around the two transitional regions where spontaneously broken weak-electromagnetic and grand-unified gauge symmetries become restored.[8]

In such a hot universe the time development of the baryon-number density $N_B(t)$ is given by

$$\frac{dN_B}{dt} = -3\frac{\dot{R}}{R}N_B + \sum_{a,b}(\Delta n_B)\langle\sigma v\rangle N_a N_b. \quad (1)$$

Here R is the cosmic scale factor; $\langle\sigma v\rangle$ is the thermal average of the reaction cross section for $a+b\to$anything times the relative velocity of a and b; Δn_B is the difference of baryon number between the final and initial states of this elementary process; N_a is the number density of a in thermal equilibrium. At the temperatures I am considering here, the energy density is dominated by highly relativistic particles and the following relation holds[2]:

$$\dot{R}/R = -\dot{T}/T = (8\pi\rho G_N/3)^{1/2}, \quad (2)$$

where ρ is the energy density

$$\rho = d_F\pi^2 T^4/15. \quad (3)$$

I use units such that the Boltzmann constant $k=1$. The effective number of degrees of freedom d_F is counted as usual, $\frac{1}{2}$ and $\frac{7}{16}$, respectively, for each boson or fermion species and spin state. To solve Eq. (1) with (2) and (3) given, it is convenient to rewrite (1) in the form

$$\frac{dF_B}{dT} = -(8\pi^3 G_N d_F/45)^{-1/2}(\tfrac{3}{8})^2 F_\gamma^{\,2}\delta, \quad (4)$$

where $\delta = \sum(\Delta n_B)\langle\sigma v\rangle$; $F_B = N_B/T^3$; $N_a = 3N_\gamma/8$; $F_\gamma = N_\gamma/T^3$ with N_γ the photon-number density. I used the fact, valid in the following example, that only massless fermions participate in baryon-number–nonconserving reactions.

To obtain a nonvanishing baryon number one must break the microscopic detailed balance (more precisely, reciprocity), because otherwise the inverse reaction would cancel the baryon number gained. This necessity of simultaneous violation of baryon-number conservation and CP or T invariance has a further consequence that the amount of generated baryon number may be severely limited. This is because the detailed balance is known[9] *not* to be broken by Born terms, and one must deal with higher-order diagrams.

As an illustration of the ideas presented above, I shall work in grand unified models based on the group SU(5), which are direct generalizations of the original Georgi-Glashow model[4] to allow more than six flavors of quarks and heavy leptons sequentially in accord with recent experimental observations.[10] Fundamental fermions are thus classified as follows:

$$\underline{5},\quad (\varphi)_{iR} = \begin{pmatrix} l_1 & q_3{}^a \\ l_2 & \end{pmatrix}_{iR} \quad (5a)$$

$$\underline{10},\quad (\psi)_{iL} = \begin{pmatrix} & q_1{}^a & \\ l_3 & & C\bar{q}_4{}^a \\ & q_2{}^a & \end{pmatrix}_{iL}, \quad (5b)$$

with several sequences ($i = 1 - n_S$). Here $(l_1 l_2)$ and $(q_1 q_2)$ form $SU(2)_W$ doublets and the others singlets; $a = 1, 2, 3$ are three colors. The baryon number is assigned as $\frac{1}{3}$ to any $SU(3)_C$ triplet and $-\frac{1}{3}$ to any antitriplet.

There are two characteristic mass scales, $m(W)$ and $m(\tilde{W})$, in this class of models with W a representative of ordinary weak bosons and with $\tilde{W}$ that of colored weak bosons. Existence of the two extremely different mass scales reflects two Higgs systems, $\tilde{H}(\underline{24})$ and (presumably several of) $\tilde{H}(\underline{5})$, being responsible for the breaking of SU(5) and $[SU(2)\otimes U(1)]_W$,[11] respectively. At the temperatures that most concern us, $m(W) \ll T \lesssim m(\tilde{W})$, the universe is effectively $[SU(2)\otimes U(1)]_W \otimes SU(3)_C$ symmetric and all fermions remain massless, which makes subsequent computations easier.

The baryon nonconservation is caused by exchange of $\tilde{W}$ coupled to fermions,

$$(g/2\sqrt{2})\sum_i[(\bar{l}_1\bar{l}_2)\gamma_\alpha(q_3{}^c)_R + \bar{l}_3\gamma_\alpha(q_2{}^c, -q_1{}^c)_L + \epsilon_{abc}(\bar{q}_1{}^a\bar{q}_2{}^a)\gamma_\alpha(C\bar{q}_4{}^b)_L]_i\begin{pmatrix}\tilde{W}_1{}^c \\ \tilde{W}_2{}^c\end{pmatrix}_\alpha + (\text{H.c.}), \quad (6)$$

and by exchange of colored Higgs $H_i(\underline{5})$ contained in the full Yukawa coupling,

$$L_Y = \tfrac{1}{2}f_{ij}\bar{\psi}_i{}^{\alpha\beta}[H_1{}^\alpha(\varphi_j{}^\beta)_R - H_1{}^\beta(\varphi_j{}^\alpha)_R] + \tfrac{1}{4}h_{ij}\epsilon_{\alpha\beta\gamma\delta\eta}\psi_i{}^{\alpha\beta}C(\psi_j{}^{\gamma\delta})_L H_2{}^\eta + (\text{H.c.}), \quad (7)$$

where the two Higgs bosons of $\underline{5}$, H_1 and H_2, may or may not coincide. The Yukawa coupling constants in (7) are related to fermion masses when $SU(2)\otimes U(1)$ is broken,

$$(f_{ij}) = (2\sqrt{2}G_F)^{1/2}\cos\zeta M_1, \quad (8a)$$

$$(h_{ij}) = (2\sqrt{2}G_F)^{1/2}\sin\zeta U^T M_2 U, \quad (8b)$$

VOLUME 41, NUMBER 5 PHYSICAL REVIEW LETTERS 31 JULY 1978

where G_F is the Fermi constant, $G_F m_N^2 \simeq 10^{-5}$; M_1 and M_2 are diagonalized mass matrices for quarks of charge $\frac{2}{3}$ and $-\frac{1}{3}$, respectively [masses of charged leptons are equal to those of $q(-\frac{1}{3})$ except for renormalization effects[12]]; U is a unitary matrix that generalizes the Cabibbo rotation for more than three flavors.[13] To allow CP nonconservation also in the Higgs sector as in the Weinberg model[14] of CP nonconservation, I introduce the complex dimensionless parameters, α and β, in propagators by

$$i\langle T(H_1{}^{\dagger a} H_2{}^a)\rangle_{q=0} = \begin{cases} \alpha/m^2(W) \text{ for } a=1,2 \\ \beta/m^2(W) \text{ for } a=3-5, \end{cases} \tag{9}$$

where q is the momentum involved in propagation. This is possible only if more than three Higgs bosons $H_i(\underline{5})$ exist with complex, trilinear and quartic couplings to $\tilde{H}(\underline{24})$.

I now calculate the quantity δ in Eq. (4) by keeping only two-body reactions. Here it is reasonable to suppose that masses of Higgs bosons are of the order of $m(W)$ for uncolored and of $m(\tilde{W})$ for colored ones and fermion masses are $\ll m(W)$, and hence $\alpha, \beta = O(1)$. Remarkably, I found after summing over all sequences that δ vanishes if α and β are real. I can actually prove that this is true to any order of perturbation. The result of this computation is particularly simple when $m(W) \ll T \ll m(\tilde{W})$. The dominant contribution to δ of (4) comes from interference of the diagrams of Figs. 1(a) and 1(b) and, after the thermal average is taken, leads to

$$\frac{\delta}{T^2} \simeq \frac{3g^2}{8\pi^2 m^4(\tilde{W})} \operatorname{Im}\beta\alpha^* [\operatorname{tr}hh^\dagger \operatorname{tr}f^2 - \operatorname{tr}hh^\dagger f^2]. \tag{10}$$

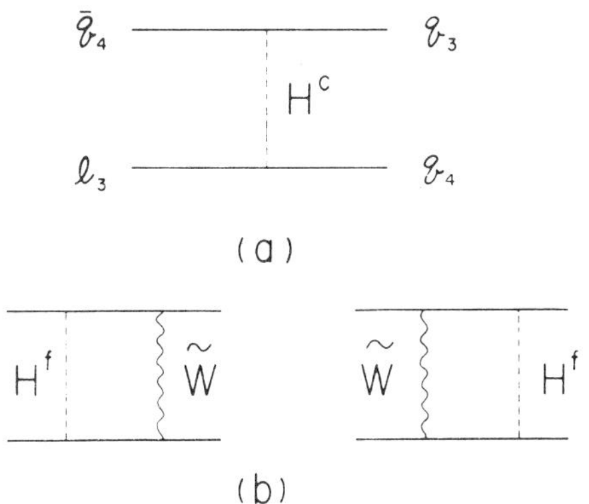

FIG. 1. Diagrams leading to generation of the baryon number. H^c (H^f) means a colored (uncolored) Higgs boson.

In this computation I ignored the Pauli exclusion effect in the final state caused by occupied thermal fermion states. Correct inclusion of this effect would not affect the result (10) drastically. Under the reasonable assumption $F_B \ll F_\gamma$ or $N_B \ll N_\gamma$, F_γ on the right-hand side of (4) may be approximated by the value at an initial temperature T_i where $N_B(T_i)=0$, and the rate equation (4) is integrated to

$$\frac{N_B(T)}{N_\gamma(T)} \simeq -(8\pi^3 G_N d_F/5)^{-1/2}(\tfrac{3}{8})^2 \frac{\delta}{T^2} N_\gamma(T_i). \tag{11}$$

During the evolution of universe from this high temperature T down to the recombination temperature (~4000°K) all fundamental constituents in the initial universe annihilate each other or go out of thermal contact, and the ratio (11) is roughly conserved to give finally the present value $(N_B/N_\gamma)_0$.

To obtain a rough quantitative idea of this ratio, I shall make a drastic extrapolation of formula (11) up to $T_i = m(\tilde{W})$. In the case of six flavors ($n_S = 3$) the best guess[12] for the parameters of the model is $g^2/4\pi = 0.022$, $m(\tilde{W}) = 2\times 10^{16}$ GeV. I also use the large difference of quark mass scales, $m(b) \gg m(s)$, $m(d)$ and $m(t) \gg m(c)$, $m(u)$, and the small mixing parameters[15] ($\lesssim 0.1$) in U of (8b). Combining (8), (10), and (11) and putting in some numerical factors, I find

$$\left(\frac{N_B}{N_\gamma}\right)_0 \approx 0.12 \frac{(d_F G_N)^{-1/2}}{m(\tilde{W})} \epsilon, \tag{12a}$$

$$\epsilon \simeq \frac{g^2}{\pi^2} G_F^2 m^2(t) m^2(b)(\sin\theta_2 + \sin\theta_3)^2 A, \tag{12b}$$

where $A = \operatorname{Im}\beta\alpha^* \sin^2\zeta\cos^2\zeta \approx O(1)$ in general. For definiteness, mixing angles are set by $\sin\theta_2 = \sin\theta_3 = 0.1$, which is consistent with present data.[15] Furthermore, $d_F = 63$ (69) with three H_i of $\underline{5}$ and real (complex) $\tilde{H}$ of $\underline{24}$, and hence

$$\left(\frac{N_B}{N_\gamma}\right)_0 \approx 2\times 10^{-9} \left[\frac{m(t)}{10 \text{ GeV}}\right]^2 \left[\frac{m(b)}{5 \text{ GeV}}\right]^2 A. \tag{13}$$

An estimate of this quantity[2] deduced from data ranges from 10^{-8} to 10^{-10}, which agrees with (13). The numerical value of (13) should not be taken too seriously because of the very crude approximations assumed, but it is hard to imagine that neglected corrections would alter (13) by more than 100.

Although my result (13) depends on the specific SU(5) model taken, the formula (12a) is presumably more general than this example provided that $\epsilon \approx (CP\text{-invariance violation parameter})/137$.

VOLUME 41, NUMBER 5 PHYSICAL REVIEW LETTERS 31 JULY 1978

To this extent the small ratio of the number densities appears to be a general consequence of grand unification in the earliest history of our universe. The possibility that this fundamental parameter of cosmology is related to those of elementary-particle physics seems intriguing and deserves much investigation.

---

[1]See e.g., T. Weeks, *High Energy Astrophysics* (Chapman and Hall, London, 1969).

[2]For a review, see S. Weinberg, *Gravitation and Cosmology* (Wiley, New York, 1972), Chap. 15.

[3]However, for a possibility of elevating the baryon number to an absolute conservation law in grand unified models, see M. Yoshimura, Progr. Theor. Phys. 58, 972 (1977); M. Abud, F. Buccella, H. Ruegg, and C. A. Savoy, Phys. Lett. 67B, 313 (1977); P. Langacker, G. Segrè, and A. Weldon, Phys. Lett. 73B, 87 (1978). Also M. Gell-Mann, P. Ramond, and R. Slansky, to be published.

[4]H. Georgi and S. L. Glashow, Phys. Rev. Lett. 32, 438 (1974); H. Fritzsch and P. Minkowski, Ann. Phys. (N.Y.) 93, 193 (1975); F. Gürsey and P. Sikivie, Phys. Rev. Lett. 36, 775 (1976); K. Inoue, A. Kakuto, and Y. Nakano, Progr. Theor. Phys. 58, 630 (1977).

[5]F. Reines and M. F. Crouch, Phys. Rev. Lett. 32, 493 (1974).

[6]H. D. Politzer, Phys. Rev. Lett. 30, 1346 (1973); D. J. Gross and F. Wilczek, Phys. Rev. Lett. 30, 1343 (1973).

[7]J. C. Collins and M. J. Perry, Phys. Rev. Lett. 34, 1353 (1975); M. B. Kislinger and P. D. Morley, Phys. Lett. 67B, 371 (1977).

[8]D. A. Kirzhnits and A. D. Linde, Phys. Lett. 42B, 471 (1972); S. Weinberg, Phys. Rev. D 9, 3357 (1974); C. Bernard, Phys. Rev. D 9, 3312 (1974); L. Dolan and R. Jackiw, Phys. Rev. D 9, 3320 (1974).

[9]See, e.g., J. M. Blatt and V. F. Weisskopf, *Theoretical Nuclear Physics* (Wiley, New York, 1952), p. 530.

[10]S. Herb *et al.*, Phys. Rev. Lett. 39, 252 (1977); W. R. Innes *et al.*, Phys. Rev. Lett. 39, 1240, 1640(E) (1977); M. L. Perl, in *Proceedings of the International Symposium on Lepton and Photon Interactions at High Energies, Hamburg, 1977,* edited by F. Gutbrod (DESY, Hamburg, Germany, 1977).

[11]S. Weinberg, Phys. Rev. Lett. 19, 1264 (1967); A. Salam, in *Elementary Particle Physics*, edited by N. Svartholm (Almquist and Wiksels, Stockholm, 1968), p. 367.

[12]H. Georgi, H. R. Quinn, and S. Weinberg, Phys. Rev. Lett. 33, 451 (1974); A. J. Buras, J. Ellis, M. K. Gaillard, and D. V. Nanopoulos, to be published.

[13]M. Kobayashi and T. Maskawa, Progr. Theor. Phys. 49, 652 (1973).

[14]S. Weinberg, Phys. Rev. Lett. 37, 657 (1976).

[15]J. Ellis, M. K. Gaillard, and D. V. Nanopoulos, Nucl. Phys. B109, 213 (1976).